Content

Why this book?

1. For the first time, you have come across a book on magic square that show you how to solve all the magic squares with just one principle.

2. It is also the first book to incorporate the principle of yin-yang in solving the magic square. In short, it relates the yin-yang principle with magic square.

3. King Fu Hsi used the MS3 to develop a new knowledge in Yin-yang Pakua or The Trigram in year 2800 BC. Later, in year 1100 BC, King Wen developed the I Ching or The Book of Change from the Trigram. Just after King Wen, his son, Duke of Chou developed the changing lines in I Ching which we now know today.

4. Now, for the first time you have the knowledge in magic square, which has a relationship with the yin-yang principle. Thus, you can try to relate the magic square and yin-yang principle to develop the new knowledge of your own. History has proved it.

Who should read this book?

This book must be read by all who have passion in mathematics irrespective of ages, that includes;

1. This book is very useful for adults who have passion in mathematics irrespective of ages.

2. This book is useful for students and teachers who have the enthusiasm in mathematics.

3. This book is useful for working class peoples who have the interest in working with numbers.

How to use this book?

1. Before you read the method of solving the magic square, please spend some time in solving magic square 3 and magic square 4.

2. Once you have solved it by yourself, continue reading the chapter one. Then read chapter two follow by chapter three. During this period, use your logic plus your creativity to solve the question put forward to you, and at the same time try to satisfy your curiosity in solving the question that might bombard you. This is to allow your mind to work while reading. This is because you need some logic and analysing power to master the magic square.

3. After mastered the **2N+1** and 4n magic squares family, try to solve magic square 6. Use the principle of symmetry.

4. If you can find the method of solving magic square 6, congratulation to you. You have somehow mastered the principle in solving the magic square by your own. If however, you feel you lack of perserverance to solve MS6, then read through chapter four. Try to master this.

5. Once you have mastered the **4N+2** magic square, try to think of algebra in play. Factoring. Just like you have used for **4N+2 = 2x (2N+1)**. Can you figure it out? If can, you have somehow mastered magic square of chapter five. Good.

6. If not, read through chapter five to know about the compound magic square. Then, a complete knowledge of magic square is in your head. Yes! your head. Congratulation!!!

1. INTRODUCTION

1.1 What is Magic Square?

Have you come across your friend asking you to solve a puzzle by arranging a number from **1** to **9** in a square of **3x3** so that the sum of each row, column or diagonal is 15? If yes, then you have already come across magic square. This is called **Magic Square 3** or in short **MS3**. Of course, as a beginner you will face difficulties to solve this. One of the solution is as shown below. Later I will show you the methodology of solving this.

8	1	6
3	5	7
4	9	2

A magic square N is defined as an arrangement of number from 1 to N^2 in a square of NxN such that the sum of each row, column or diagonal is the same.

This sum denotes the value of magic square. In the later section, we will see how to pre-calculate this value before solving the magic square. Let us look at the solution for **MS3** again, as illustrated below.

			d2
8	1	6	*r1*
3	5	7	*r2*
4	9	2	*r3*
c1	*c2*	*c3*	*d1*

Notice that the sum of the each row is

row 1:- $r1 = 8 + 1 + 6 = 15$
row 2:- $r2 = 3 + 5 + 7 = 15$
row 3:- $r3 = 4 + 9 + 2 = 15$

each column is

column 1:- $c1 = 8 + 3 + 4 = 15$
column 2:- $c2 = 1 + 5 + 9 = 15$
column 3:- $c3 = 6 + 7 + 2 = 15$

or each diagonal is

diagonal 1:- $d1 = 8 + 5 + 2 = 15$
diagonal 2:- $d2 = 4 + 5 + 6 = 15$

Notice that each rows, columns or diagonals has the same value of 15. Thus, **MS3** has a value of 15. Take note that the number **3** of **MS3** denotes the size of the magic square.

1.2 Order of Magic Square

Order of magic square is just refers to the square that makes up the magic square. For instance, magic square of 3x3 is of order 3, and magic square 30x30 is of order 30. And it is named as magic square 3 and magic square 30 respectively. In short, and going forward we just name it as MS3 or MS30.

1.3   Value of Magic Square

We have seen the value of the **MS3** is 15 after solving the magic square. How do we get this value before solving the magic square? The value is defined as the ratio of the sum of all the numbers of the magic square over its size. Let us calculate the value of **MS3**.

Value of MS3
= Sum of numbers from 1 to 9 divides by size
= $(1 + 2 + 3 + 4 + 5 + 6 + 7 + 8 + 9) / 3$
= $45 / 3$
= 15

Simple indeed for MS3. In general, the value of a Magic Square N is defined as

the ratio of sum of all the number from 1 to N^2 over the size N

Making use of this definition, we obtained the value of the magic square as shown in the table below:

Magic Square	Value	Magic Square	Value
3	15	17	2465
4	34	18	2925

5	65	19	3439
6	111	20	4010
7	175	21	4641
8	260	22	5335
9	369	23	6095
10	505	24	6924
11	671	25	7825
12	870	26	8801
13	1105	27	9855
14	1379	28	10990
15	1695	29	12209
16	2056	30	13515

Table 1: The value of Magic Square from 3 to 30

Question 1

As the first practice, can you get the value ot the magic as shown in the Table 1 above.

Hint: Please use the formula $N(N+1)/2$ to calculate the sum of the all the numbers

1.4   Category of Magic Square

Magic Square can be categorised into 3 main groups, namely

a. $2N + 1$; $N = 1, 2, 3, \ldots$
     (3, 5, 7, . . .)

This is an odd symmetry magic square.

b.   4**N**   ; **N** = 1, 2, 3, . . .
      (4, 8, 12, . . .)
This is an even symmetry magic square.

c.   4**N** + 2   ; **N** = 1, 2, 3, . . .
      (6, 10, 14, . . .)
This is an even asymmetry magic square.

Later we will see that the methodology of Magic Square evolves from the symmetry of magic square.

1.5 Brief History of Magic Square

Magic Square is believes to exist since 2800 BC ago in China in the time of King Fu Hsi. There is a story to relate how the king happenned to know about the magic square. One day, while strolling along the river bank, the king happened to notice a strange tortoise. When he picked up the tortoise, he noticed that there is an arrangement of number below the tortoise shell. The arrangement is the magic square of 3 or MS3 as shown below.

4	9	2
3	5	7
8	1	6

He was curious and spent his time in the arrangement. Finally, he managed to come out with a method in solving this arrangement, the magic square 3. It is so happened that the method formulated by him can be used to solve the other odd magic square. However, there is no written note to state that Fu Hsi managed to develop the higher odd magic square.

The magic square 3 then became the foundation for Fu Hsi to develop a new kind of knowledge, the trigram or yin-yang pakua, which still exists until now. Later King Wen developed the Book of Change or I Ching from the principle of the pakua. King Wen son, Duke of Chou, continued his father works on the I Ching and later came out with the knowledge of the changing lines. Thus, magic square 3 becomes the foundation to the development of pakua and then I Ching.

Magic square 3 also been used by the Feng Shui masters of the Flying Stars school. This knowledge of Feng Shui still exists today.

I believe that you will be able to use this knowledge learned from this book to develop a new kind of knowledge in other fields. Maybe, it becomes not a magic square but magic to you to come out with new knowledge and invention, in other fields, who knows.

1.6 Benefit of Magic Square

a. Once you are being asked to solve the MS3, of course you will crack your head for 20 or 30 minutes trying to arrange the nine number. In the process of solving it, indirectly you are practising your mental strength which are

(a) logic
(b) creativity
(c) positive thinking
(d) perseverance
(e) courage

in order. Of course, solving MS3, let say about 30 minutes, you can only increase your mental strength slightly. For this, you need a harder challenge, that means a longer time spend. Therefore, this book which put forward the solution of MS3 to MS18 will surely put your brain at work, all the time, so as to strengthen it.

b. Once you have already master the magic square method, and with your creativity you achieve through the process of mastering it, you will be able to formulate a new kind of method in solving the magic square. Also, the mental strength that you have acquired through solving the magic square can be used in other fields such as engineering, science, technology, design, etc. for the benefits of oneself and mankind.

c. The book also teaches you of using computer spreadsheet programs to solve the magic square. With this process, you will be able to master the spreadsheet.

d. After you have masters magic square until 18, with the curiosity and willingness to know the unknown, you can spend your time wisely in solving the higher magic square, that is higher than 18. Of course, during your leisure time.

1.7 Principle of Magic Square

Magic square can be categorised into 3 types, namely

a. Odd magic square, or in mathematical form can be written as **2N+1**, where **N =1, 2, 3**, ... Putting the value of **N**, we obtain the MS3, MS5, MS7, ...

b. Multiple of 4 magic square, and in mathematical form can be written as **4N**, where **N = 1, 2, 3**, ... Putting the value of **N**, we obtain the MS4, MS8, MS12, ...

c. **4N+2**, where **N = 1, 2, 3**, … . Putting the value of **N**, we obtain the MS6, MS10, MS14, … .

The categorisation of the magic square is based on the principle of symmetry. In order to prove the symmetry principle, put **N = 0**, thus we have

a. For **2N+1 = 2(0) + 1 = 1**. It is the magic square 1 or MS1. It is solves automatically; row, column and diagonal. The extra MS1 becomes the backbone of type **2N+1** magic square. The extra MS1 defines that the symmetry is along the middle row or column. For better illustration, let us see the simplest one in the family, MS3, below. (This symmetry applies to all members in the family).

1	2	3
4	5	6
7	8	9

The symmetry row is {4,5,6} and the symmetry column is {2,5,8}. It consists of cell MS1

b. For **4N = 4(0) = 0**. It is just 0. With 0, no need to solve. The symmetry is along the middle lines of the row and column. For better illustration, let us wee the simplest one in the family, MS4, below. (This symmetry applies to all members in the family).

	y-axis		
1	2	3	4
5	6	7	8
9	10	11	12
13	14	15	16

x-axis

The symmetries are denote by x-axis and y-axis for horinzontal and vertical lines respectively

c. For **4N+2 = 4(0) + 2 = 2**. It is the MS2. Notice that MS2 cannot be solved. It can either be solved by row, by column or by diagonal as shown below.

1	2
3	4

diagonal solved

1	4
3	2

row solved

1	2
4	3

column solved

There is no symmetry. Thus, magic square of type 4N+2 is asymmetry magic square, whereas magic square of type 2N+1 is odd symmetry magic square and 4N is even symmetry magics square. The method to solve the magic square depends on the category.

2. ODD MAGIC SQUARE (2N +1)

In the earlier chapter, we noticed that the method of solving the magic square depends on the category of the magic square. This is because the method lies in the symmetry principle of the magic square. In order for us to solve this magic square **2N+1**, it is best to look at the simplest magic square of this type, that is MS3. The method of solving MS3 is as follows.

2.1 Magic Square 3 where N = 1

This is an odd magic square. It is a symmetry magic square. In order for us to understand this magic square, let us look at its simplest form, that is, **MS3**. If we start with the starting arrangement as shown below, we notice that the symmetry lines are highlighted.

1	2	3
4	5	6
7	8	9

Row **2** consists of **4**, **5** and **6** whereas Column **2** consists of **2**, **5** and **8** are the symmetry row and column respectively. Making use of this symmetry let us solved **MS3**.

Case 1:- Row clockwise - Column anticlockwise

Step 1

Starting arrangement

1	2	3
4	5	6
7	8	9

Step 2

Rotate clockwise along the symmetry row, {4,5,6}.

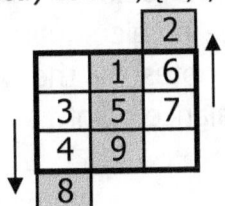

Step 3

*Put **3** & **7** back to the original square (red).*

3	1	2
4	5	6
8	9	7

Step 4

Rotate anticlockwise along the symmetry column, { 1,5,9}.

Step 5

*Put **8** & **2** back to the original square (red).*

8	1	6
3	5	7
4	9	2

Solution

*There you are the solution for **MS3***

			15
8	1	6	15
3	5	7	15
4	9	2	15
15	15	15	15

17

2.2 Magic Square 3 Yin-Yang Principle

In step 1, everything evolve from the basic order. In step 2, there is a rotating force in the clockwise direction pushing the order along the invariant line, symmetry row, the immovable. In step 3, the imbalance force take becomes temporary balance within the immovable red square and ·takes its position. In step 4, there is a rotating force in the anticlockwise direction pushing the order along the invariant line, symmetry column. the immovable. And lastly, the force become balance within the immovable red square and takes its position, thus the magic square of 3 comes into existence. This is denotes by the value of the magic square, 15 the balance point.

In solving this magic square, we do make use of the immovable and the movable entities. The rotation of the forces are in the opposite direction, that is clockwise and anticlockwise, that is yin-yang. Thus, MS3 comes into existence through the imbalance forces, yin & yang, imposing onto the orderlity, the starting arrangement. Using the same understanding, we can rotate anticlockwise first then clockwise to get the result of MS3.

Case 2:- Column anticlockwise - Row clockwise

Step 1

Starting arrangement

1	2	3
4	5	6
7	8	9

Step 2

Rotate anticlockwise along the symmetry column, {2,5,8}.

		3
	2	6
1	5	9
4	8	
7		

Step 3

Put 3 & 7 back to the original square (red).

7	2	6
1	5	9
4	8	3

Step 4

Rotate clockwise along the symmetry row, { 1,5,9}.

		7	2	6
	1	5	9	
4	8	3		

Step 5

Put 4 & 6 back to the original square (red).

6	7	2
1	5	9
8	3	4

Solution

There you are another solution for MS3.

			15
6	7	2	15
1	5	9	15
8	3	4	15
15	15	15	15

2.3 Magic Square 5 where N = 2

Similarly, we solve Magic Square 5 as above methodology, clockwise then anticlockwise direction.

Case 1:- Row clockwise - Column anticlockwise
Step 1   **Step 2**

1	2	3	4	5
6	7	8	9	10
11	12	13	14	15
16	17	18	19	20
21	22	23	24	25

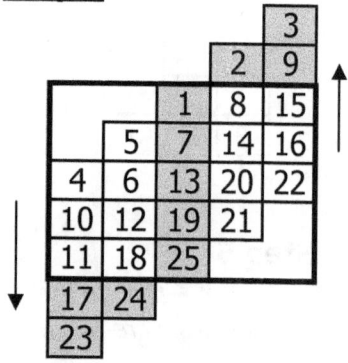

Step 3   **Step 4**

4	5	1	2	3
10	6	7	8	9
11	12	13	14	15
17	18	19	20	16
23	24	25	21	22

Step 5   **Solution 1 for MS5**

17	24	1	8	15
23	5	7	14	16
4	6	13	20	22
10	12	19	21	3
11	18	25	2	9

17	24	1	8	15	65
23	5	7	14	16	65
4	6	13	20	22	65
10	12	19	21	3	65
11	18	25	2	9	65
65	*65*	*65*	*65*	*65*	*65*

Question 2

Can you find the invariant square, row and column?

Question 3

Can you find the pattern of the solution 1 for MS5? Later, I will show you the pattern methodology.

Case 2:- Column anticlockwise - Row clockwise

Step 1

1	2	3	4	5
6	7	8	9	10
11	12	13	14	15
16	17	18	19	20
21	22	23	24	25

Step 2

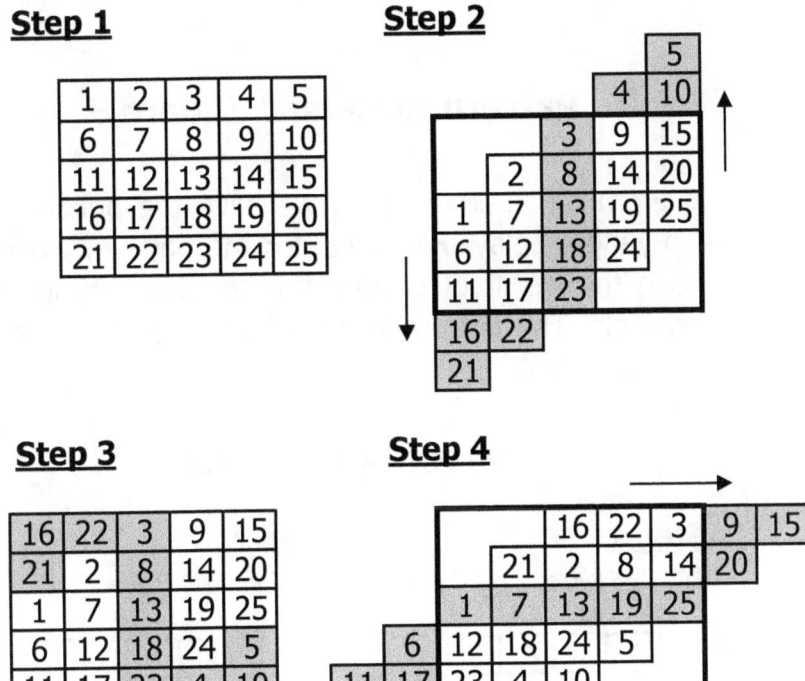

Step 3

16	22	3	9	15
21	2	8	14	20
1	7	13	19	25
6	12	18	24	5
11	17	23	4	10

Step 4

Step 5

9	15	16	22	3
20	21	2	8	14
1	7	13	19	25
12	18	24	5	6
23	4	10	11	17

Solution 2 for MS5

9	15	16	22	3	65
20	21	2	8	14	65
1	7	13	19	25	65
12	18	24	5	6	65
23	4	10	11	17	65
65	65	65	65	65	65

Question 4

Can you find the pattern of the solution 1 for MS5? Later, I will show you the pattern methodology.

Pattern for Magic Square 5

From the methodology, we have the principle of yin-yang at play. However, for the pattern, we just have the result of getting the solution without the principle. The pattern and methodology for Question 3 is as below.

NE Movement Pattern

Step 1
Put 1 as shown

Step 2
Put 2 in the NE direction.

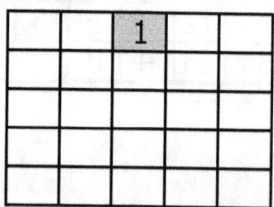

Step 3
Put 3 & 4 in the NE direction

		1		
4				
			3	
		2		

Step 4
Put 5 in the NE direction

		1		
	5			
4				
			3	
		2		

Step 5
Put 6 below 5 as 1 already occupies the cell.

		1		
	5			
4	6			
			3	
		2		

Step 6
Put 7, 8 & 9 in the NE direction.

		1	8	
	5	7		
4	6			
			3	
		2	9	

Step 7

Put 10 in the NE direction.

		1	8	
	5	7		
4	6			
10			3	
		2	9	

Step 8
Put 11 below 10 as 6 already occupies the cell.

		1	8	
	5	7		
4	6			
10			3	
11		2	9	

Step 9

Put 12, 13, 14 & 15 in the NE direction.

		1	8	15
	5	7	14	
4	6	13		
10	12			3
11			2	9

Step 10

Put 16 below 15 as there is no cell for 16.

		1	8	15
	5	7	14	16
4	6	13		
10	12			3
11			2	9

Step 11

Repeat Step 6, 7, & 8.

17	24	1	8	15
23	5	7	14	16
4	6	13	20	22
10	12	19	21	3
11	18	25	2	9

Step 12

The pattern as shown

17	24	1	8	15
23	5	7	14	16
4	6	13	20	22
10	12	19	21	3
11	18	25	2	9

The pattern and method for Question 4 is as below.

Horse Move Movement Pattern

Step 1

Put 1, 2 & 3 in the horse move.

				3
		2		
1				

Step 2

Put 4 & 5 in the horse move.

				3
		2		
1				
			5	
		4		

Step 3

When reach multiple of 5, put 6 beside 5.

				3
		2		
1				
			5	6
	4			

Step 4

Put 7, 8, 9 & 10 in the horse move.

9				3
		2	8	
1	7			
			5	6
	4	10		

Step 5

When reach multiple of 5, put 11 beside 10

9				3
		2	8	
1	7			
			5	6
	4	10	11	

Step 6

Put 12, 13, 14 & 15 in the horse move.

9	15			3
		2	8	14
1	7	13		
12			5	6
	4	10	11	

Step 7

When reach multiple of 5, put 16 beside 15.

9	15	16		3
		2	8	14
1	7	13		
12			5	6
	4	10	11	

Step 8

Put 17, 18, 19 & 20 in the horse move.

9	15	16		3
20		2	8	14
1	7	13	19	
12	18		5	6
	4	10	11	17

Step 9

When reach multiple of 5, put 21 beside 20.

9	15	16		3
20	21	2	8	14
1	7	13	19	
12	18		5	6
	4	10	11	17

Step 10

Put 22, 23, 24 & 25 in the horse move.

9	15	16	22	3
20	21	2	8	14
1	7	13	19	25
12	18	24	5	6
23	4	10	11	17

Question 5

Can you get the solution for Magic Square 7 using the pattern method?

2.4   Magic Square 7 where N = 3

Column anticlockwise - Row clockwise
Step 1

1	2	3	4	5	6	7
8	9	10	11	12	13	14
15	16	17	18	19	20	21
22	23	24	25	26	27	28
29	30	31	32	33	34	35
36	37	38	39	40	41	42
43	44	45	46	47	48	49

Step 2

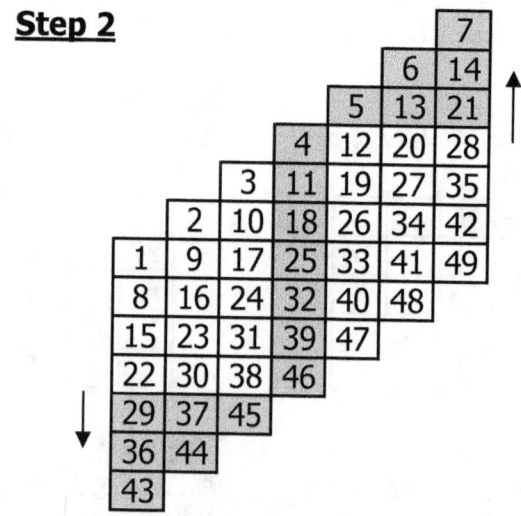

Step 3

29	37	45	4	12	20	28
36	44	3	11	19	27	35
43	2	10	18	26	34	42
1	9	17	25	33	41	49
8	16	24	32	40	48	7
15	23	31	39	47	6	14
22	30	38	46	5	13	21

Step 4

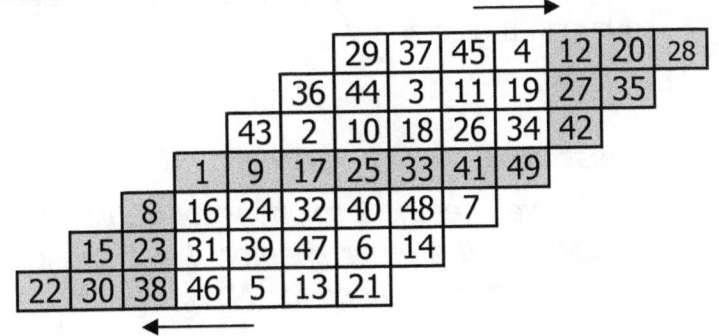

Step 5

12	20	28	29	37	45	4
27	35	36	44	3	11	19
42	43	2	10	18	26	34
1	9	17	25	33	41	49
16	24	32	40	48	7	8
31	39	47	6	14	15	23
46	5	13	21	22	30	38

Solution to MS7

							175
12	20	28	29	37	45	4	175
27	35	36	44	3	11	19	175
42	43	2	10	18	26	34	175
1	9	17	25	33	41	49	175
16	24	32	40	48	7	8	175
31	39	47	6	14	15	23	175
46	5	13	21	22	30	38	175

175 175 175 175 175 175 175 175

By mastering the methodology of solving Magic Square of 2N + 1 you can solve the higher magic square in the category.

Question 6

*Can you get the 2 solutions for Magic Square **9** ? Later in the other chapter, I will show how to obtain more solution for **MS9**.*

2.5 Special Case for Magic Square 3.

We have seen the methodology of solving magic square 3 using the column and row symmetry. However for MS3 there is another symmetry that is the diagonal symmetry as shown below. This methodology is only applicable to tbe basic unit in the family, that is MS3.

1	2	3
4	5	6
7	8	9

Notice that the diagonal symmetry are {1, 5 , 9} & {3, 5, 7} as highlighted in yellow cells. Thus, using the same methodology we can solve the Magic Square 3 along the diagonal symmetry. Let us see below.

{7,5,3} clockwise - {8,5,2} anticlockwise
Step 1
Starting arrangement.

1	2	3
4	5	6
7	8	9

Step 2
Notice that cells 2, 4, 8 & 6 move 1 cell whereas 1 and 9 move 2 cells along the invariant diagonal {7,5,3} in a clockwise direction.

Step 3
Putting back to the original square (red).

8	4	3
9	5	1
7	6	2

Step 4
Notice that 1, 4, 9 & 6 move 1 cell whereas 3 and 7 move 2 cells along the invariant diagonal (8,5,2) in an anticlockwise direction.

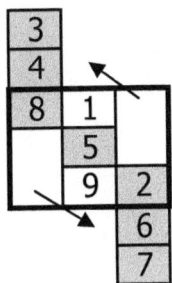

Step 5
Putting back to the original square (red).

8	1	6
3	5	7
4	9	2

Solution to MS 3

8	1	6	*15*
3	5	7	*15*
4	9	2	*15*

   15 15 15

2.6 Solving Magic Square 3 with Excel

Row clockwise - column Anticlockwise

Step 1
Starting arrangement

Step 2
Rotate clockwise along ivariant row, {4,5,6}.

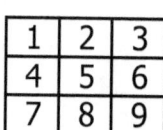

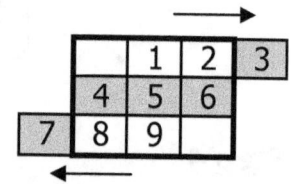

Step 3

Move {3} and {7} to original square (red).

3	1	2
4	5	6
8	9	7

Step 4

Rotate anticlockwise along invariant column {1,5,9}.

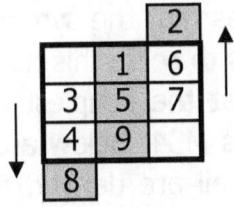

Step 5

Move {2} and {8} to original square (red).

8	1	6
3	5	7
4	9	2

Solution

			15
8	1	6	15
3	5	7	15
4	9	2	15
15	15	15	15

3. EVEN MAGIC SQUARE (4N)

In this chapter, we will learn how to solve an even magic square. From the symmetry test, that is putting **N=0**, we have **4N = 4(0) = 0**. This implies that the symmetry is just a line. In order for us to solve this magic square **4N**, it is best to look at the simplest magic square of this type, that is MS4. Below are the 6 methods of solving MS4, all are derived from the symmetry of the magic square.

3.1 Magic Square 4 where N = 1

This is an even magic square. It is a symmetry magic square. In order for us to understand this magic square, let us look at its simplest form, that is, **MS4**. If we start with the starting arrangement as shown below, we notice that the symmetry are the horizontal and the vertical red lines.

1	2	3	4
5	6	7	8
9	10	11	12
13	14	15	16

If we put in some notation, we have as below diagram.

1	2	3	4
5	6	7	8
9	10	11	12
13	14	15	16

r1 r2 r3 r4 (rows); c1 c2 c3 c4 (columns)

Thus, we have 4 rows and 4 columns. Notice that the mirror image for **r1** is **r4** and the mirror image for **r2** is **r3**. So are the mirror image for **c1** and **c2** are **c4** and **c3** respectively. By using the concepts of symmetry and the mirror image we have the methodology of solving this family of magic square. Let us solve **MS4**.

Method 1

Step 1
Starting arrangement.

1	2	3	4
5	6	7	8
9	10	11	12
13	14	15	16

r1 r2 r3 r4 (rows); c1 c2 c3 c4 (columns)

Step 2
*Fixed **c1** and its image **c4**, then rotate **c2** and its image **c3** about the horizontal symmetry by 180^0.*

1	14	15	4
5	10	11	8
9	6	7	12
13	2	3	16

r1 r2 r3 r4 (rows); c1 c2 c3 c4 (columns)

Step 3

*Fixed r **1** and its image r **4**, then rotate r **2** and its image r**3** about the vertical symmetry by 180⁰.*

	c1	c2	c3	c4
r1	1	14	15	4
r2	8	11	10	5
r3	12	7	6	9
r4	13	2	3	16

Solution 1 for MS4

1	14	15	4	34
8	11	10	5	34
12	7	6	9	34
13	2	3	16	34
34	34	34	34	34

3.2 Magic Square 4 Yin-Yang Principle

In step 1, everything evolve from the basic order. In step 2, we separate orderlity into the movable (c2 & c3) and the immovable (c1 & c4). The imbalance force about the horizontal move the movable by 180^0 and get its temporary position. Then, in step 3, we separate temporary orderlity into the movable (r2 & r3) and the immovable (r1 & r4). The imbalance about the vertical move the movable by 180^0 and get its position. Thus, MS4 comes into existence. This is denotes by the value of the magic square, 34, the balance.

In solving this magic square, we do make use of the immovable and the movable entities. The rotation of the forces are in the opposite direction, that is about the horizontal and about the vertical, that is **yin-yang**. Thus, MS4 comes into existence through the imbalance forces, yin & yang, imposing onto the orderlity, the starting arrangement. Using the same understanding, interchange the movable and the immovable among the columns and rows accordingly.

Method 2

Step 1

Starting arrangement.

	c1	c2	c3	c4
r1	1	2	3	4
r2	5	6	7	8
r3	9	10	11	12
r4	13	14	15	16

Step 2

*Fixed **c2** and its image **c3**, then rotate **c1** and its image **c4** about the horizontal symmetry by 180^0.*

	c1	c2	c3	c4
r1	13	2	3	16
r2	9	6	7	12
r3	5	10	11	8
r4	1	14	15	4

Step 3

*Fixed r **2** and its image r **3**, then rotate r **1** and its image r**4** about the vertical symmetry by 180^0.*

r1	16	3	2	13
r2	9	6	7	12
r3	5	10	11	8
r4	4	15	14	1
	c1	c2	c3	c4

Solution 2 for MS4

				34
16	3	2	13	34
9	6	7	12	34
5	10	11	8	34
4	15	14	1	34
34	34	34	34	34

Using the same methodology, we can solve Magic Square as below table to obtain the variation solution for MS4.

Variation	Step 1				Step 2			
	Fixed		Rotate		Fixed		Rotate	
1	c1	c4	c2	c3	r1	r4	r2	r3
2	c1	c4	c2	c3	r2	r3	r1	r4
3	c2	c3	c1	c4	r1	r4	r2	r3
4	c2	c3	c1	c4	r2	r3	r1	r4
5	r1	r4	r2	r3	c1	c4	c2	c3
6	r2	r3	r1	r4	c1	c4	c2	c3
7	r1	r4	r2	r3	c2	c3	c1	c4
8	r2	r3	r1	r4	c2	c3	c1	c4

Table 2: The Variation to the solution of Magic Square 4

Question 7

Can you obtain the solutions as per Table 2?

3.3 Magic Square 8 where N = 2

Similarly, we can use the above methodology to solve Magic Square 8.

Method 1

Step 1

Starting arrangement

	c1	c2	c3	c4	c5	c6	c7	c8
r1	1	2	3	4	5	6	7	8
r2	9	10	11	12	13	14	15	16
r3	17	18	19	20	21	22	23	24
r4	25	26	27	28	29	30	31	32
r5	33	34	35	36	37	38	39	40
r6	41	42	43	44	45	46	47	48
r7	49	50	51	52	53	54	55	56
r8	57	58	59	60	61	62	63	64

Step 2

Fixed c1 & c2 and their images c8 & c7 respectively.
Rotate c3 & c4 and their images c6 & c5 respectively about the horizontal line by 180^0 .

	c1	c2	c3	c4	c5	c6	c7	c8
r1	1	2	59	60	61	62	7	8
r2	9	10	51	52	53	54	15	16
r3	17	18	43	44	45	46	23	24
r4	25	26	35	36	37	38	31	32
r5	33	34	27	28	29	30	39	40
r6	41	42	19	20	21	22	47	48
r7	49	50	11	12	13	14	55	56
r8	57	58	3	4	5	6	63	64

Step 3

Fixed r3 & r4 and their images r6 & r5 respectively. Rotate r1 & r2 and their images r8 & r7 respectively about the vertical line by 180°.

	c1	c2	c3	c4	c5	c6	c7	c8
r1	8	7	62	61	60	59	2	1
r2	16	15	54	53	52	51	10	9
r3	17	18	43	44	45	46	23	24
r4	25	26	35	36	37	38	31	32
r5	33	34	27	28	29	30	39	40
r6	41	42	19	20	21	22	47	48
r7	56	55	14	13	12	11	50	49
r8	64	63	6	5	4	3	58	57

Note: Notice that we interchange the immovable and the movable at will.

Solution for MS8

								Sum
8	7	62	61	60	59	2	1	260
16	15	54	53	52	51	10	9	260
17	18	43	44	45	46	23	24	260
25	26	35	36	37	38	31	32	260
33	34	27	28	29	30	39	40	260
41	42	19	20	21	22	47	48	260
56	55	14	13	12	11	50	49	260
64	63	6	5	4	3	58	57	260
260	260	260	260	260	260	260	260	260

Method 2

Step 1

Starting arrangement

	c1	c2	c3	c4	c5	c6	c7	c8
r1	1	2	3	4	5	6	7	8
r2	9	10	11	12	13	14	15	16
r3	17	18	19	20	21	22	23	24
r4	25	26	27	28	29	30	31	32
r5	33	34	35	36	37	38	39	40
r6	41	42	43	44	45	46	47	48
r7	49	50	51	52	53	54	55	56
r8	57	58	59	60	61	62	63	64

Step 2

Fixed c2 & c4 and their images c7 & c5 respectively.
Rotate c1 & c3 and their images c8 & c6 respectively about the horizontal line by 180^0.

	c1	c2	c3	c4	c5	c6	c7	c8
r1	57	2	59	4	5	62	7	64
r2	49	10	51	12	13	54	15	56
r3	41	18	43	20	21	46	23	48
r4	33	26	35	28	29	38	31	40
r5	25	34	27	36	37	30	39	32
r6	17	42	19	44	45	22	47	24
r7	9	50	11	52	53	14	55	16
r8	1	58	3	60	61	6	63	8

Step 3

Fixed r2 & r3 and their images r7 & r6 respectively. Rotate r1 & r4 and their images r8 & r5 respectively about the vertical line by 180⁰ .

	c1	c2	c3	c4	c5	c6	c7	c8
r1	64	7	62	5	4	59	2	57
r2	49	10	51	12	13	54	15	56
r3	41	18	43	20	21	46	23	48
r4	40	31	38	29	28	35	26	33
r5	32	39	30	37	36	27	34	25
r6	17	42	19	44	45	22	47	24
r7	9	50	11	52	53	14	55	16
r8	8	63	6	61	60	3	58	1

Note: Notice that we interchange the immovable and the movable at will.

Solution for MS8

								260
64	7	62	5	4	59	2	57	260
49	10	51	12	13	54	15	56	260
41	18	43	20	21	46	23	48	260
40	31	38	29	28	35	26	33	260
32	39	30	37	36	27	34	25	260
17	42	19	44	45	22	47	24	260
9	50	11	52	53	14	55	16	260
8	63	6	61	60	3	58	1	260
260	260	260	260	260	260	260	260	260

Question 8

Can you obtain variation for Magic Square 8 as per Table 2? Then, try to obtain the solution.

Question 9

Can you obtain solution for Magic Square 12?

With the methodology above, you can solve any magic square in this family.

3.4 Special Cases for Magic Square 4

We have seen the methodology of solving magic square 4. However for MS4 as a basic unit in the family there a few ways of obtaining the other variation besides the variation in Table 2. This methodology is only applicable to tbe basic unit in the family, that is MS4.

Diagonal Symmetry

Step 1

Starting arrangement.

1	2	3	4
5	6	7	8
9	10	11	12
13	14	15	16

Step 2

Rotate diagonal (1, 6, 11, 16) about the origin by 180^0.

16	2	3	4
5	11	7	8
9	10	6	12
13	14	15	1

Step 3

Rotate diagonal (4, 7, 10, 13) about the origin by 180^0.

16	2	3	13
5	11	10	8
9	7	6	12
4	14	15	1

Fixed Diagonal

Step 1

Starting arrangement. Fixed diagonals {1,6,11,16} and {13,10,7,4}.

1	2	3	4
5	6	7	8
9	10	11	12
13	14	15	16

Step 2

Rotate sides (2, 3, 8, 12, 15, 14, 9, 5) by 180^0 about the origin.

1	15	14	4
12	6	7	9
8	10	11	5
13	3	2	16

3.5 Special Case for Magic Square 8

We have seen the methodology of solving magic square 8. This special case is obtained by rotating along the fixed center point rather than along the vertical or horizontal line.

Step 1

Starting arrangement

	c1	c2	c3	c4	c5	c6	c7	c8
r1	1	2	3	4	5	6	7	8
r2	9	10	11	12	13	14	15	16
r3	17	18	19	20	21	22	23	24
r4	25	26	27	28	29	30	31	32
r5	33	34	35	36	37	38	39	40
r6	41	42	43	44	45	46	47	48
r7	49	50	51	52	53	54	55	56
r8	57	58	59	60	61	62	63	64

Step 2

Fixed c1 & c2 and their images c8 & c7 respectively.
Rotate c3 & c4 and their images c6 & c5 respectively about
the origin by 180⁰.

	c1	c2	c3	c4	c5	c6	c7	c8
r1	1	2	62	61	60	59	7	8
r2	9	10	54	53	52	51	15	16
r3	17	18	46	45	44	43	23	24
r4	25	26	38	37	36	35	31	32
r5	33	34	30	29	28	27	39	40
r6	41	42	22	21	20	19	47	48
r7	49	50	14	13	12	11	55	56
r8	57	58	6	5	4	3	63	64

Step 3

Fixed r1 & r2 and their images r8 & r7 respectively. Rotate
r3 & r4 and their images r6 & r5 respectively about the
origin by 180⁰.

	c1	c2	c3	c4	c5	c6	c7	c8
r1	1	2	62	61	60	59	7	8
r2	9	10	54	53	52	51	15	16
r3	48	47	19	20	21	22	42	41
r4	40	39	27	28	29	30	34	33
r5	32	31	35	36	37	38	26	25
r6	24	23	43	44	45	46	18	17
r7	49	50	14	13	12	11	55	56
r8	57	58	6	5	4	3	63	64

Solution for MS8

								260
1	2	62	61	60	59	7	8	260
9	10	54	53	52	51	15	16	260
48	47	19	20	21	22	42	41	260
40	39	27	28	29	30	34	33	260
32	31	35	36	37	38	26	25	260
24	23	43	44	45	46	18	17	260
49	50	14	13	12	11	55	56	260
57	58	6	5	4	3	63	64	260
260	260	260	260	260	260	260	260	260

Note:

This method can be applied to any member of the family of 4N. You can try to solve the MS4 and MS12.

3.6 Magic Square 4 by Induction

We have seen above that there are few methods of solving **MS4**. In this section, we would like to solve **MS4** just by induction method. Let us understand further the regarding MS4. Look at one of the solution of MS4. It has 4 properties as shown below.

Properties 1

Balance Quadrant

1	14	15	4
8	11	10	5
12	7	6	9
13	2	3	16

Properties 2

Balance Center

1	14	15	4
8	11	10	5
12	7	6	9
13	2	3	16

Properties 3
Balance Corner

1	14	15	4
8	11	10	5
12	7	6	9
13	2	3	16

Properties 4
Balance Side

1	14	15	4
8	11	10	5
12	7	6	9
13	2	3	16

With these properties MS4 can be induced just one reflection or multiple reflection on the symmetry. Let us see for few variation as shown below.

Variation 1

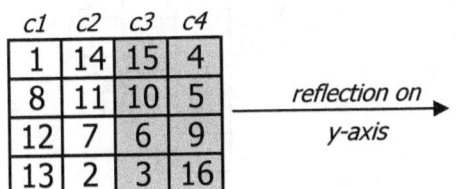

c1	c2	c3	c4
1	14	15	4
8	11	10	5
12	7	6	9
13	2	3	16

reflection on y-axis

c3	c4	c1	c2
15	4	1	14
10	5	8	11
6	9	12	7
3	16	13	2

Variation 2

r1	1	14	15	4
r2	8	11	10	5
r3	12	7	6	9
r4	13	2	3	16

reflection on y-axis

r3	12	7	6	9
r4	13	2	3	16
r1	1	14	15	4
r2	8	11	10	5

Variation 3

1	14	15	4
8	11	10	5
12	7	6	9
13	2	3	16

reflection on xy-axis

6	9	15	4
3	16	10	5
12	7	1	14
13	2	8	11

Question 10

Making use of the induction method, try to obtain as many solution for MS4.

3.7 Solving Magic Square 4 with Excel

Step 1
Starting arrangement.

Step 2
Move the shaded column as above

1	2	3	4
5	6	7	8
9	10	11	12
13	14	15	16

1			4
5			8
9			12
13	2	3	16
	6	7	
	10	11	
	14	15	

Step 3
Move {{6,7} as shown.

Step 4
Move {10,11} as shown.

1			4
5			8
9	6	7	12
13	2	3	16

10	11
14	15

1			4
5	10	11	8
9	6	7	12
13	2	3	16

14	15

Step 5

Move {14,15} as shown.

1	14	15	4
5	10	11	8
9	6	7	12
13	2	3	16

Step 6

Move the shaded row as shown.

1	14	15	4			
			5	10	11	8
			9	6	7	12
13	2	3	16			

Step 7

Move {10,6} as shown.

1	14	15	4
		10	5
		6	9
13	2	3	16

Step 8

Move { 11,7} as shown.

1	14	15	4
	11	10	5
	7	6	9
13	2	3	16

8
12

Step 9

Move {8,12} as shown.

1	14	15	4
8	11	10	5
12	7	6	9
13	2	3	16

Practise hard. Practice make perfect.

4. ASYMETRY MAGIC SQUARE (4N + 2)

This is an even magic square. It is an asymmetry magic square. This means that it has no symmetry line. Thus, the methodology of symmetry cannot be performed as in the earlier families, 2N + 1 and 4N. Thus to understand further this magic square, we need to look at the basic cell of the magic square.

From the principle of symmetry, the magic square of type **4N+2** is an asymmetry magic square. Of course we cannot solve the magic square by making use of symmetry axis directly. In order to solve the asymmetry magic square we need to do a kind of transformation. Let us look at the mathematical form of this magic square, that is **4N+2**. From algebra, **4N+2 = 2x(2N+1)**. The expression **2x(2N+1)** consists of the odd symmetry magic square, that is **2N+1**, but with a factor 2. This factor 2 is MS2. Earlier we have shown that it can only be solved either by row, by column or by diagonal, but not all. That is why the asymmetry term is used. In short the expression is

$$4N+2 = 2x(2N+1)$$

is a transformation tool to transform asymmetry magic square **4N+2** to compound magic square of asymmetrical 2 and symmetrical **2N+1**. By considering the asymmetrical factor 2 as invariant, we have a symmetry magic square **2N+1**, on temporary basis only. In other words, the expression **2x(2N+1)** can be said to be a magic square of type **2N+1**, however each cell is a magic square 2 or MS2, instead of a single cell or MS1. Knowing this concept of transformation, thus, the method of solving the magic square 4N+2 can is as follow.

Methodology for 4N + 2

1. Transform the magic square **4N+2** to magic square symmetry magic square of **2N+1** consisting of asymmetry MS2.

2. Solve the symmetry magic square as magic square **2N+1,** (remember the cell is now MS2 instead of MS1). Once this is done, the symmetry part of the magic square is solved.

3. Still remember, for magic square **2N+1** there is a symmetry row and symmetry column. But in this case, the symmetry row and symmetry column are of MS2 instead of a single cell or MS1. Thus, we need to perform interchanging of cell (MS1) within the MS2 for this symmetry row and symmetry column. Once this is done, the symmetry column and the symmetry row are solved.

4. Now we have left with the asymmetry MS2 at the outside of the symmetry column and symmetry row. In order to solve this, we need to remember the pattern shown below.

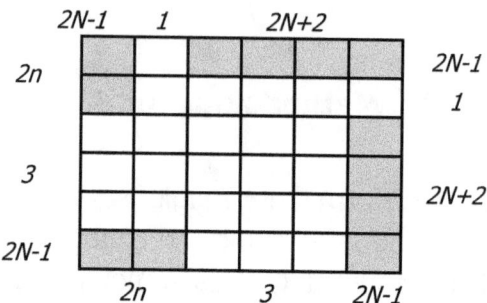

*The term N refers to the term N of the expression **4N+2**. Notice that all the sides have a sum of **4N+2**. Try to add them up. Therefore, by interchanging the cells (MS1) within MS2 of the coloured area, you will obtain the solution for magic square **4N+2**. For the row shaded area, you interchange the cells column-wise (yin) just within the MS2 only, and for the column shaded area, you interchange the cells row-wise (yang) just within the MS2 only.*

4.1 Magic Square 6 where N = 1

Let us solve the simplest form of this family that is MS6 where N = 1. MS6 can be express as 2 x MS3. This means that it is just a MS3 but the building cell is MS2. Thus to solve this MS3, I expect that you have already mastered how to solve the odd magic square or MS3.

Step 1
Starting Arrangement

1	2	3	4	5	6
7	8	9	10	11	12
13	14	15	16	17	18
19	20	21	22	23	24
25	26	27	28	29	30
31	32	33	34	35	36

Step 2
Transforming into MS3 with cell MS2.

1	2	3	4	5	6
7	8	9	10	11	12
13	14	15	16	17	18
19	20	21	22	23	24
25	26	27	28	29	30
31	32	33	34	35	36

Step 3 to Step 6 just follows the methodology of solving MS3.

Step 3

		1	2	3	4	5	6
		7	8	9	10	11	12
	13	14	15	16	17	18	
	19	20	21	22	23	24	
25	26	27	28	29	30		
31	32	33	34	35	36		

Step 4

5	6	1	2	3	4
11	12	7	8	9	10
13	14	15	16	17	18
19	20	21	22	23	24
27	28	29	30	25	26
33	34	35	36	31	32

51

Step 5

				3	4
				9	10
		1	2	17	18
		7	8	23	24
5	6	15	16	25	26
11	12	21	22	31	32
13	14	29	30		
19	20	35	36		
27	28				
33	34				

Step 6

27	28	1	2	17	18
33	34	7	8	23	24
5	6	15	16	25	26
11	12	21	22	31	32
13	14	29	30	3	4
19	20	35	36	9	10

From step 1 to step 6, we have completely solved the symmetrical part of the magic square. So we are left with the asymmetrical part of the magic square. If we will to calculate the value of MS2, we have is 5. Let us look at MS 2 again in details.

1	2
3	4

diagonal balance as it has a value of 5, but the column and row are not balance

1	4
2	3

row balance as it has a value of 5, but the column and diagonal are not balance

1	2
4	3

column balance as it has a value of 5, but the row and diagonal are not balance

With this unbalance nature of MS2, thus the magic square of type 4N + 2 is not a symmetry magic square. However, there is either diagonal, row or column balance nature in the MS2 magic square. Making use of this balance nature, we will then tuning the unbalance nature of MS2 in the solution of the symmetrical part of the magic square as in Step 6.

Step 7

For the symmetrical column of MS2, we just interchange vertically the cells in the MS2. Thus, we have 1 with 7, 2 with 8, 15 with 21, 16 with 22, 29 with 35 and 30 with 36

27	28	7	8	17	18
33	34	1	2	23	24
5	6	21	22	25	26
11	12	15	16	31	32
13	14	35	36	3	4
19	20	29	30	9	10

Step 8

For the symmetrical row of MS2, we just interchange horizontally the cells in the MS2. Thus, we have 5 interchange with 6, 12 with 11, 22 with 21, 16 with 15, 26 with 25 and 32 with 31

27	28	7	8	17	18
33	34	1	2	23	24
6	5	22	21	26	25
12	11	16	15	32	31
13	14	35	36	3	4
19	20	29	30	9	10

Step 9

At last we interchange 8 with 7, 18 with 17, 33 with 27, 20 with 19, 4 with 10 and 25 with 31 all within the cell of MS2.

33	28	8	7	18	17
27	34	1	2	23	24
6	5	22	21	26	31
12	11	16	15	32	25
13	14	35	36	3	10
20	19	29	30	9	4

Solution for MS6

						111
33	28	8	7	18	17	111
27	34	1	2	23	24	111
6	5	22	21	26	31	111
12	11	16	15	32	25	111
13	14	35	36	3	10	111
20	19	29	30	9	4	111
111	111	111	111	111	111	111

There you are the solution of MS6. This is quite tedious compares to the earlier once because we need to solve both the symmetrical and the asymmetrical parts. The pattern as shown in Step 9 above will be explained further when we are solving MS10.

Question 11

How many solution of MS6 can you obtain using this method? I got more than 10.

4.2   Magic Square 6 Yin-Yang Principle

In step 1, everything evolve from the basic order. However, this order is not symmetrical. In step 2, the orderlity transform itself to symmetrical magic square, MS3 with MS2 cells. In step 3, there is a rotating force in the clockwise direction pushing the order along the invariant line, symmetry row, the immovable. In step 4, the imbalance force take becomes temporary balance within the immovable red square and takes its position. In step 5, there is a rotating force in the anticlockwise direction pushing the order along the invariant line, symmetry column. the immovable. In step 6, the force become balance within the immovable square and takes its position, and the symmetrical part of the magic square is balance.

However, within the MS2, the forces are still imbalance. In step 7, there is an interchange of cells within the symmetrical column thus balance the forces of movement there. In step 8, there is an interchange of cells within the symmetrical row thus balance the forces of movement there. Notice that step 7 and step 8 are concept of yin-yang within the symmetrical column and row. In step 9, we balance the side cells according to the pattern as highlighted in green. And every forces are balance, thus MS6 comes into existence.

In solving this magic square, we solve the symmetrical part (yin), and then the asymmetrical part (yang). For the symmetrical part, the yin-yang forces acted according to the forces as explained in the MS3. For the asymmetrical part, the yin-yang just acted within the MS2. To solve this, we first tackle the immovable part, column and row. Once the force within this immovable is okay, we then solve the movable part, that is the cells at the sides. In trying to solve this, it follow a unique pattern as highlighted in green and blue as in step 9. This pattern will be explained further when we solved MS10.

4.3 Magic Square 10 where N = 2

Similarly, we solve Magic Square 10 as above methodology.

Step 1

Starting arrangement.

1	2	3	4	5	6	7	8	9	10
11	12	13	14	15	16	17	18	19	20
21	22	23	24	25	26	27	28	29	30
31	32	33	34	35	36	37	38	39	40
41	42	43	44	45	46	47	48	49	50
51	52	53	54	55	56	57	58	59	60
61	62	63	64	65	66	67	68	69	70
71	72	73	74	75	76	77	78	79	80
81	82	83	84	85	86	87	88	89	90
91	92	93	94	95	96	97	98	99	100

Step 2

Transforming into MS5 with cell MS2

1	2	3	4	5	6	7	8	9	10
11	12	13	14	15	16	17	18	19	20
21	22	23	24	25	26	27	28	29	30
31	32	33	34	35	36	37	38	39	40
41	42	43	44	45	46	47	48	49	50
51	52	53	54	55	56	57	58	59	60
61	62	63	64	65	66	67	68	69	70
71	72	73	74	75	76	77	78	79	80
81	82	83	84	85	86	87	88	89	90
91	92	93	94	95	96	97	98	99	100

For Step 3 to Step 8, it is just a magic square of 5. However, for Step 3 to Step 6, it done partially for top and lower parts of the invariant row.

Step 3

Rotate right part first (done partially for large magic square).

1	2	3	4	5	6	7	8	9	10
11	12	13	14	15	16	17	18	19	20
21	22	23	24	25	26	27	28	29	30
31	32	33	34	35	36	37	38	39	40
41	42	43	44	45	46	47	48	49	50
51	52	53	54	55	56	57	58	59	60
61	62	63	64	65	66	67	68	69	70
71	72	73	74	75	76	77	78	79	80
81	82	83	84	85	86	87	88	89	90
91	92	93	94	95	96	97	98	99	100

Step 4

Putting back to the original square.

7	8	9	10	1	2	3	4	5	6
17	18	19	20	11	12	13	14	15	16
29	30	21	22	23	24	25	26	27	28
39	40	31	32	33	34	35	36	37	38
41	42	43	44	45	46	47	48	49	50
51	52	53	54	55	56	57	58	59	60
61	62	63	64	65	66	67	68	69	70
71	72	73	74	75	76	77	78	79	80
81	82	83	84	85	86	87	88	89	90
91	92	93	94	95	96	97	98	99	100

Step 5

Rotate the left part (remember the direction).

7	8	9	10	1	2	3	4	5	6
17	18	19	20	11	12	13	14	15	16
29	30	21	22	23	24	25	26	27	28
39	40	31	32	33	34	35	36	37	38
41	42	43	44	45	46	47	48	49	50
51	52	53	54	55	56	57	58	59	60
61	62	63	64	65	66	67	68	69	70
71	72	73	74	75	76	77	78	79	80
81	82	83	84	85	86	87	88	89	90
91	92	93	94	95	96	97	98	99	100

Step 6

Putting back to the original square.

7	8	9	10	1	2	3	4	5	6
17	18	19	20	11	12	13	14	15	16
29	30	21	22	23	24	25	26	27	28
39	40	31	32	33	34	35	36	37	38
41	42	43	44	45	46	47	48	49	50
51	52	53	54	55	56	57	58	59	60
63	64	65	66	67	68	69	70	61	62
73	74	75	76	77	78	79	80	71	72
85	86	87	88	89	90	81	82	83	84
95	96	97	98	99	100	91	92	93	94

Step 7

Rotate anticlockwise along the invariant column.

								5	6
								15	16
						3	4	27	28
						13	14	37	38
				1	2	25	26	49	50
				11	12	35	36	59	60
		9	10	23	24	47	48	61	62
		19	20	33	34	57	58	71	72
7	8	21	22	45	46	69	70	83	84
17	18	31	32	55	56	79	80	93	94
29	30	43	44	67	68	81	82		
39	40	53	54	77	78	91	92		
41	42	65	66	89	90				
51	52	75	76	99	100				
63	64	87	88						
73	74	97	98						
85	86								
95	96								

Step 8

Putting back to the original square.

63	64	87	88	1	2	25	26	49	50
73	74	97	98	11	12	35	36	59	60
85	86	9	10	23	24	47	48	61	62
95	96	19	20	33	34	57	58	71	72
7	8	21	22	45	46	69	70	83	84
17	18	31	32	55	56	79	80	93	94
29	30	43	44	67	68	81	82	5	6
39	40	53	54	77	78	91	92	15	16
41	42	65	66	89	90	3	4	27	28
51	52	75	76	99	100	13	14	37	38

At this point, the symmetrical parts of MS10 is solved, MS5. The next step is to solve the asymmetrical part, MS2.

Step 9

For the symmetrical column, we just interchange column-wise the cells within the MS2. Thus, we have 1 with 11, 2 with 12, 23 with 33, 24 with 34, 45 with 55, 46 with 56, 67 with 77, 68 with 78, 89 with 99 and 90 with 100.

63	64	87	88	11	12	25	26	49	50
73	74	97	98	1	2	35	36	59	60
85	86	9	10	33	34	47	48	61	62
95	96	19	20	23	24	57	58	71	72
7	8	21	22	55	56	69	70	83	84
17	18	31	32	45	46	79	80	93	94
29	30	43	44	77	78	81	82	5	6
39	40	53	54	67	68	91	92	15	16
41	42	65	66	99	100	3	4	27	28
51	52	75	76	89	90	13	14	37	38

Step 10

Similarly, for the symmetrical row, we just interchange row-wise for the cells within the MS2. Thus, we have 7 with 8, 17 with 18, 21 with 22, 31 with 32, 55 with 56, 45 with 46, 69 with 70, 79 with 80, 83 with 84 and 93 with 94.

63	64	87	88	11	12	25	26	49	50
73	74	97	98	1	2	35	36	59	60
85	86	9	10	33	34	47	48	61	62
95	96	19	20	23	24	57	58	71	72
8	7	22	21	56	55	70	69	84	83
18	17	32	31	46	45	80	79	94	93
29	30	43	44	77	78	81	82	5	6
39	40	53	54	67	68	91	92	15	16
41	42	65	66	99	100	3	4	27	28
51	52	75	76	89	90	13	14	37	38

Step 11

Lastly, we interchange cells within the cell MS2.

73	74	97	88	12	11	26	25	50	49
63	64	87	98	2	1	36	35	60	59
95	96	19	10	34	33	48	47	62	61
85	86	9	20	23	24	57	58	71	72
8	7	22	21	56	55	70	79	94	93
18	17	32	31	46	45	80	69	84	83
29	30	43	44	77	78	81	92	15	16
40	39	54	53	67	68	91	82	5	6
42	41	66	65	99	100	3	14	37	38
52	51	76	75	89	90	13	4	27	28

Solution for MS10

73	74	97	88	12	11	26	25	50	49	505
63	64	87	98	2	1	36	35	60	59	505
95	96	19	10	34	33	48	47	62	61	505
85	86	9	20	23	24	57	58	71	72	505
8	7	22	21	56	55	70	79	94	93	505
18	17	32	31	46	45	80	69	84	83	505
29	30	43	44	77	78	81	92	15	16	505
40	39	54	53	67	68	91	82	5	6	505
42	41	66	65	99	100	3	14	37	38	505
52	51	76	75	89	90	13	4	27	28	505
505	505	505	505	505	505	505	505	505	505	505

4.4 The Pattern for Magic Square 4N + 2

In solving MS6 and MS10, we make use of the pattern in the last step. Luckily this pattern has a unique relationship with the formula, and thus help us to solve the magic square easily. Let us look at the Figure below.

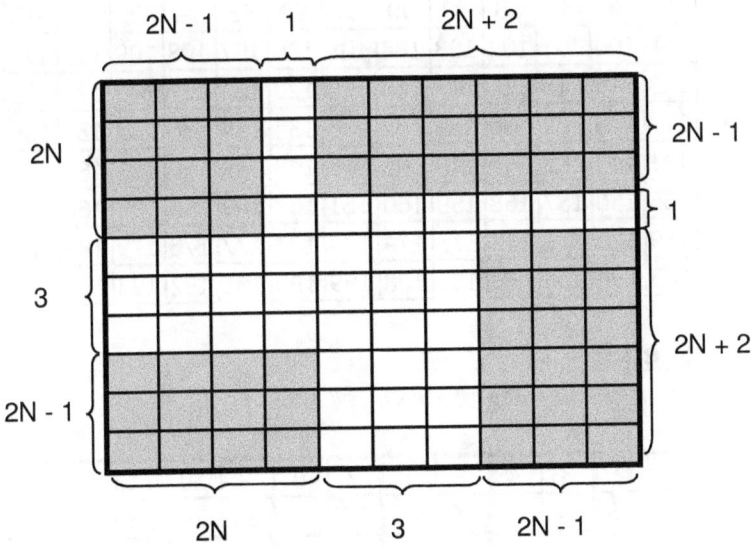

Pattern of 4N + 2:- The respective sides denotes the number of cells to be interchange in the MS2. The value of N refers to the value of N in the magic square 4N + 1.

Making use of the pattern, let us solve the magic square 14 or MS14.

Step 1

Starting arrangement.

1	2	3	4	5	6	7	8	9	10	11	12	13	14
15	16	17	18	19	20	21	22	23	24	25	26	27	28
29	30	31	32	33	34	35	36	37	38	39	40	41	42
43	44	45	46	47	48	49	50	51	52	53	54	55	56
57	58	59	60	61	62	63	64	65	66	67	68	69	70
71	72	73	74	75	76	77	78	79	80	81	82	83	84
85	86	87	88	89	90	91	92	93	94	95	96	97	98
99	100	101	102	103	104	105	106	107	108	109	110	111	112
113	114	115	116	117	118	119	120	121	122	123	124	125	126
127	128	129	130	131	132	133	134	135	136	137	138	139	140
141	142	143	144	145	146	147	148	149	150	151	152	153	154
155	156	157	158	159	160	161	162	163	164	165	166	167	168
169	170	171	172	173	174	175	176	177	178	179	180	181	182
183	184	185	186	187	188	189	190	191	192	193	194	195	196

Step 2

Transforming into MS7 with cell MS2.

1	2	3	4	5	6	7	8	9	10	11	12	13	14
15	16	17	18	19	20	21	22	23	24	25	26	27	28
29	30	31	32	33	34	35	36	37	38	39	40	41	42
43	44	45	46	47	48	49	50	51	52	53	54	55	56
57	58	59	60	61	62	63	64	65	66	67	68	69	70
71	72	73	74	75	76	77	78	79	80	81	82	83	84
85	86	87	88	89	90	91	92	93	94	95	96	97	98
99	100	101	102	103	104	105	106	107	108	109	110	111	112
113	114	115	116	117	118	119	120	121	122	123	124	125	126
127	128	129	130	131	132	133	134	135	136	137	138	139	140
141	142	143	144	145	146	147	148	149	150	151	152	153	154
155	156	157	158	159	160	161	162	163	164	165	166	167	168
169	170	171	172	173	174	175	176	177	178	179	180	181	182
183	184	185	186	187	188	189	190	191	192	193	194	195	196

Step 3

We break the normal step into 2 parts, the right part and the left part. Thus, the right part is

1	2	3	4	5	6	7	8	9	10	11	12	13	14
15	16	17	18	19	20	21	22	23	24	25	26	27	28
29	30	31	32	33	34	35	36	37	38	39	40	41	42
43	44	45	46	47	48	49	50	51	52	53	54	55	56
57	58	59	60	61	62	63	64	65	66	67	68	69	70
71	72	73	74	75	76	77	78	79	80	81	82	83	84
85	86	87	88	89	90	91	92	93	94	95	96	97	98
99	100	101	102	103	104	105	106	107	108	109	110	111	112
113	114	115	116	117	118	119	120	121	122	123	124	125	126
127	128	129	130	131	132	133	134	135	136	137	138	139	140
141	142	143	144	145	146	147	148	149	150	151	152	153	154
155	156	157	158	159	160	161	162	163	164	165	166	167	168
169	170	171	172	173	174	175	176	177	178	179	180	181	182
183	184	185	186	187	188	189	190	191	192	193	194	195	196

Step 4

Putting back into the original square.

9	10	11	12	13	14	1	2	3	4	5	6	7	8
23	24	25	26	27	28	15	16	17	18	19	20	21	22
39	40	41	42	29	30	31	32	33	34	35	36	37	38
53	54	55	56	43	44	45	46	47	48	49	50	51	52
69	70	57	58	59	60	61	62	63	64	65	66	67	68
83	84	71	72	73	74	75	76	77	78	79	80	81	82
85	86	87	88	89	90	91	92	93	94	95	96	97	98
99	100	101	102	103	104	105	106	107	108	109	110	111	112
113	114	115	116	117	118	119	120	121	122	123	124	125	126
127	128	129	130	131	132	133	134	135	136	137	138	139	140
141	142	143	144	145	146	147	148	149	150	151	152	153	154
155	156	157	158	159	160	161	162	163	164	165	166	167	168
169	170	171	172	173	174	175	176	177	178	179	180	181	182
183	184	185	186	187	188	189	190	191	192	193	194	195	196

Step 5
And left part.

						9	10	11	12	13	14	1	2	3	4	5	6	7	8
						23	24	25	26	27	28	15	16	17	18	19	20	21	22
						39	40	41	42	29	30	31	32	33	34	35	36	37	38
						53	54	55	56	43	44	45	46	47	48	49	50	51	52
						69	70	57	58	59	60	61	62	63	64	65	66	67	68
						83	84	71	72	73	74	75	76	77	78	79	80	81	82
						85	86	87	88	89	90	91	92	93	94	95	96	97	98
						99	100	101	102	103	104	105	106	107	108	109	110	111	112
				113	114	115	116	117	118	119	120	121	122	123	124	125	126		
				127	128	129	130	131	132	133	134	135	136	137	138	139	140		
		141	142	143	144	145	146	147	148	149	150	151	152	153	154				
		155	156	157	158	159	160	161	162	163	164	165	166	167	168				
169	170	171	172	173	174	175	176	177	178	179	180	181	182						
183	184	185	186	187	188	189	190	191	192	193	194	195	196						

Step 6
Putting back into the original square.

9	10	11	12	13	14	1	2	3	4	5	6	7	8
23	24	25	26	27	28	15	16	17	18	19	20	21	22
39	40	41	42	29	30	31	32	33	34	35	36	37	38
53	54	55	56	43	44	45	46	47	48	49	50	51	52
69	70	57	58	59	60	61	62	63	64	65	66	67	68
83	84	71	72	73	74	75	76	77	78	79	80	81	82
85	86	87	88	89	90	91	92	93	94	95	96	97	98
99	100	101	102	103	104	105	106	107	108	109	110	111	112
115	116	117	118	119	120	121	122	123	124	125	126	113	114
129	130	131	132	133	134	135	136	137	138	139	140	127	128
145	146	147	148	149	150	151	152	153	154	141	142	143	144
159	160	161	162	163	164	165	166	167	168	155	156	157	158
175	176	177	178	179	180	181	182	169	170	171	172	173	174
189	190	191	192	193	194	195	196	183	184	185	186	187	188

Step 7

Rotate anti-clockwise along the invariant column.

												7	8
												21	22
									5	6		37	38
									19	20		51	52
							3	4	35	36		67	68
							17	18	49	50		81	82
					1	2	33	34	65	66		97	98
					15	16	47	48	79	80		111	112
			13	14	31	32	63	64	95	96		113	114
			27	28	45	46	77	78	109	110		127	128
	11	12	29	30	61	62	93	94	125	126		143	144
	25	26	43	44	75	76	107	108	139	140		157	158
9	10	41	42	59	60	91	92	123	124	141	142	173	174
23	24	55	56	73	74	105	106	137	138	155	156	187	188
39	40	57	58	89	90	121	122	153	154	171	172		
53	54	71	72	103	104	135	136	167	168	185	186		
69	70	87	88	119	120	151	152	169	170				
83	84	101	102	133	134	165	166	183	184				
85	86	117	118	149	150	181	182						
99	100	131	132	163	164	195	196						
115	116	147	148	179	180								
129	130	161	162	193	194								
145	146	177	178										
159	160	191	192										
175	176												
189	190												

Step 8

Putting back to the original square, thus solving the symmetrical part, MS7.

115	116	147	148	179	180	1	2	33	34	65	66	97	98
129	130	161	162	193	194	15	16	47	48	79	80	111	112
145	146	177	178	13	14	31	32	63	64	95	96	113	114
159	160	191	192	27	28	45	46	77	78	109	110	127	128
175	176	11	12	29	30	61	62	93	94	125	126	143	144
189	190	25	26	43	44	75	76	107	108	139	140	157	158
9	10	41	42	59	60	91	92	123	124	141	142	173	174
23	24	55	56	73	74	105	106	137	138	155	156	187	188
39	40	57	58	89	90	121	122	153	154	171	172	7	8
53	54	71	72	103	104	135	136	167	168	185	186	21	22
69	70	87	88	119	120	151	152	169	170	5	6	37	38
83	84	101	102	133	134	165	166	183	184	19	20	51	52
85	86	117	118	149	150	181	182	3	4	35	36	67	68
99	100	131	132	163	164	195	196	17	18	49	50	81	82

Step 9

Interchange column-wise the cells within MS2 of the symmetry column.

115	116	147	148	179	180	15	16	33	34	65	66	97	98
129	130	161	162	193	194	1	2	47	48	79	80	111	112
145	146	177	178	13	14	45	46	63	64	95	96	113	114
159	160	191	192	27	28	31	32	77	78	109	110	127	128
175	176	11	12	29	30	75	76	93	94	125	126	143	144
189	190	25	26	43	44	61	62	107	108	139	140	157	158
9	10	41	42	59	60	105	106	123	124	141	142	173	174
23	24	55	56	73	74	91	92	137	138	155	156	187	188
39	40	57	58	89	90	135	136	153	154	171	172	7	8
53	54	71	72	103	104	121	122	167	168	185	186	21	22
69	70	87	88	119	120	165	166	169	170	5	6	37	38
83	84	101	102	133	134	151	152	183	184	19	20	51	52
85	86	117	118	149	150	195	196	3	4	35	36	67	68
99	100	131	132	163	164	181	182	17	18	49	50	81	82

Step 10

Interchange row-wise the cells within the MS2 of the symmetry row.

115	116	147	148	179	180	15	16	33	34	65	66	97	98
129	130	161	162	193	194	1	2	47	48	79	80	111	112
145	146	177	178	13	14	45	46	63	64	95	96	113	114
159	160	191	192	27	28	31	32	77	78	109	110	127	128
175	176	11	12	29	30	75	76	93	94	125	126	143	144
189	190	25	26	43	44	61	62	107	108	139	140	157	158
10	9	42	41	60	59	106	105	124	123	142	141	174	173
24	23	56	55	74	73	92	91	138	137	156	155	188	187
39	40	57	58	89	90	135	136	153	154	171	172	7	8
53	54	71	72	103	104	121	122	167	168	185	186	21	22
69	70	87	88	119	120	165	166	169	170	5	6	37	38
83	84	101	102	133	134	151	152	183	184	19	20	51	52
85	86	117	118	149	150	195	196	3	4	35	36	67	68
99	100	131	132	163	164	181	182	17	18	49	50	81	82

Step 11

Interchange the cells within MS2 of the pattern for 4N + 2.

129	130	161	162	193	180	16	15	34	33	66	65	98	97
115	116	147	148	179	194	2	1	48	47	80	79	112	111
159	160	191	192	27	14	46	45	64	63	96	95	114	113
145	146	177	178	13	28	32	31	78	77	110	109	128	127
189	190	25	26	43	30	76	75	94	93	126	125	144	143
175	176	11	12	29	44	61	62	107	108	139	140	157	158
10	9	42	41	60	59	106	105	124	137	156	155	188	187
24	23	56	55	74	73	92	91	138	123	142	141	174	173
39	40	57	58	89	90	135	136	153	168	185	186	21	22
54	53	72	71	104	103	121	122	167	154	171	172	7	8
70	69	88	87	120	119	165	166	169	184	19	20	51	52
84	83	102	101	134	133	151	152	183	170	5	6	37	38
86	85	118	117	150	149	195	196	3	18	49	50	81	82
100	99	132	131	164	163	181	182	17	4	35	36	67	68

Solution for MS14

1379

129	130	161	162	193	180	16	15	34	33	66	65	98	97	1379
115	116	147	148	179	194	2	1	48	47	80	79	112	111	1379
159	160	191	192	27	14	46	45	64	63	96	95	114	113	1379
145	146	177	178	13	28	32	31	78	77	110	109	128	127	1379
189	190	25	26	43	30	76	75	94	93	126	125	144	143	1379
175	176	11	12	29	44	61	62	107	108	139	140	157	158	1379
10	9	42	41	60	59	106	105	124	137	156	155	188	187	1379
24	23	56	55	74	73	92	91	138	123	142	141	174	173	1379
39	40	57	58	89	90	135	136	153	168	185	186	21	22	1379
54	53	72	71	104	103	121	122	167	154	171	172	7	8	1379
70	69	88	87	120	119	165	166	169	184	19	20	51	52	1379
84	83	102	101	134	133	151	152	183	170	5	6	37	38	1379
86	85	118	117	150	149	195	196	3	18	49	50	81	82	1379
100	99	132	131	164	163	181	182	17	4	35	36	67	68	1379

1379 1379 1379 1379 1379 1379 1379 1379 1379 1379 1379 1379 1379 1379 1379

After you have completed this section, you have mastered how to solve magic square of 4N + 2. And also all the magic square. Congratulation!

4.5 Solving Magic Square 6 with Excel

Step 1

Starting arrangement.

1	2	3	4	5	6
7	8	9	10	11	12
13	14	15	16	17	18
19	20	21	22	23	24
25	26	27	28	29	30
31	32	33	34	35	36

Step 2

Rotate clockwise along the invariant row.

		1	2	3	4	5	6
		7	8	9	10	11	12
	13	14	15	16	17	18	
	19	20	21	22	23	24	
25	26	27	28	29	30		
31	32	33	34	35	36		

Step 3

Putting back to the original square.

5	6	1	2	3	4
11	12	7	8	9	10
13	14	15	16	17	18
19	20	21	22	23	24
27	28	29	30	25	26
33	34	35	36	31	32

Step 4

Rotate anticlockwise along the invariant column.

				3	4
				9	10
		1	2	17	18
		7	8	23	24
5	6	15	16	25	26
11	12	21	22	31	32
13	14	29	30		
19	20	35	36		
27	28				
33	34				

Step 5

Putting back to the original square.

27	28	1	2	17	18
33	34	7	8	23	24
5	6	15	16	25	26
11	12	21	22	31	32
13	14	29	30	3	4
19	20	35	36	9	10

Step 6

Move the symmetry column by one row downward as shown.

27	28			17	18
33	34	1	2	23	24
5	6	7	8	25	26
11	12	15	16	31	32
13	14	21	22	3	4
19	20	29	30	9	10
		35	36		

Step 7

Move {7,8} as shown.

27	28	7	8	17	18
33	34	1	2	23	24
5	6			25	26
11	12	15	16	31	32
13	14	21	22	3	4
19	20	29	30	9	10
		35	36		

Step 8

Move {21,22} as shown.

27	28	7	8	17	18
33	34	1	2	23	24
5	6	21	22	25	26
11	12	15	16	31	32
13	14			3	4
19	20	29	30	9	10
		35	36		

Step 9

Move {35,36} as shown.

27	28	7	8	17	18
33	34	1	2	23	24
5	6	21	22	25	26
11	12	15	16	31	32
13	14	35	36	3	4
19	20	29	30	9	10

Step 10

Move the symmetry row by one column to the right as shown.

27	28	7	8	17	18	
33	34	1	2	23	24	
	5	6	21	22	25	26
	11	12	15	16	31	32
13	14	35	36	3	4	
19	20	29	30	9	10	

Step 11
Move {6,5} as shown.

27	28	7	8	17	18	
33	34	1	2	23	24	
6	5		21	22	25	26
12	11		15	16	31	32
13	14	35	36	3	4	
19	20	29	30	9	10	

Step 12
Move {22,16} as shown.

27	28	7	8	17	18	
33	34	1	2	23	24	
6	5	22	21		25	26
12	11	16	15		31	32
13	14	35	36	3	4	
19	20	29	30	9	10	

Step 13
Move {26,32} as shown.

27	28	7	8	17	18
33	34	1	2	23	24
6	5	22	21	26	25
12	11	16	15	32	31
13	14	35	36	3	4
19	20	29	30	9	10

Step 14

Move the shaded block as above.

27						
33	28		7	8	17	18
	34	1	2	23	24	
6	5	22	21	26		
12	11	16	15	32	25	
13	14	35	36	3	31	
19	20		29	30	9	4
						10

Step 15

Move {27}, {19}, {31} and {8} to respective MS2.

33	28	8	7		17	18
27	34	1	2	23	24
6	5	22	21	26	31
12	11	16	15	32	25
13	14	35	36	3	
20	19	29	30	9	4
					10

Step 16

Move {10} and {18} as shown.

33	28	8	7	18	17
27	34	1	2	23	24
6	5	22	21	26	31
12	11	16	15	32	25
13	14	35	36	3	10
20	19	29	30	9	4

You have solved the MS6 with Excel.

Try to use Excel to solve MS10 and MS14 in order to master the tool. Then you can use Excel to solve the higher magic square which will be tedious to write the same set of numbers over and over again.

5. COMPOUNDED MAGIC SQUARE

5.1 What is Compounded Magic Square?

How this compounded magic square comes about? Why we need to know about this compounded magic square as we can solve all the magic square using the methodology put forward in the earlier chapters? From the earlier chapters, we can solve magic with limited variation as put forward by the methodology. With this compounded magic square, we can create more variation in solving the magic square.

The compounded magic square comes from the way we solve the asymmetry magic square, 4N + 2. From this, we simplified the equation to 2(2N + 1) which can be stated as the odd magic square with cell MS2. Since we can have a cell of MS2, why not we have a cell of MS3 or MS4 and so on. With this induction, thus the compounded magic square is formed, and the methodology of solving this magic square is just the same as what we have done for the asymmetry magic square. However, the difference lies in that the compounded magic square have the cell of magic square which is higher than MS2, thus can be solved perfectly as what we have solved in the earlier chapters.

The smallest compounded magic square is MS9. From mathematics, 9 = 3 x 3. In this notation, we can express magic square 9 into magic square 3 with the cell of MS3. This is like transforming the magic square 6 to 2 x 3. Let see below how we solved magic square 9 using the compounded methodology.

Magic Square 9 = 3x3

Step 1
Starting arrangement

1	2	3	4	5	6	7	8	9
10	11	12	13	14	15	16	17	18
19	20	21	22	23	24	25	26	27
28	29	30	31	32	33	34	35	36
37	38	39	40	41	42	43	44	45
46	47	48	49	50	51	52	53	54
55	56	57	58	59	60	61	62	63
64	65	66	67	68	69	70	71	72
73	74	75	76	77	78	79	80	81

Step 2
Transforming MS9 into MS3 with cell of MS3

1	2	3	4	5	6	7	8	9
10	11	12	13	14	15	16	17	18
19	20	21	22	23	24	25	26	27
28	29	30	31	32	33	34	35	36
37	38	39	40	41	42	43	44	45
46	47	48	49	50	51	52	53	54
55	56	57	58	59	60	61	62	63
64	65	66	67	68	69	70	71	72
73	74	75	76	77	78	79	80	81

Step 3

Solving MS3 along symmetrical row of MS3 of cells MS3

			1	2	3	4	5	6	7	8	9			
			10	11	12	13	14	15	16	17	18			
			19	20	21	22	23	24	25	26	27			
		28	29	30	31	32	33	34	35	36				
		37	38	39	40	41	42	43	44	45				
		46	47	48	49	50	51	52	53	54				
55	56	57	58	59	60	61	62	63						
64	65	66	67	68	69	70	71	72						
73	74	75	76	77	78	79	80	81						

Step 4

The first temporary balance of MS3

7	8	9	1	2	3	4	5	6
16	17	18	10	11	12	13	14	15
25	26	27	19	20	21	22	23	24
28	29	30	31	32	33	34	35	36
37	38	39	40	41	42	43	44	45
46	47	48	49	50	51	52	53	54
58	59	60	61	62	63	55	56	57
67	68	69	70	71	72	64	65	66
76	77	78	79	80	81	73	74	75

Step 5

Solving MS3 along symmetrical column of MS3 of cells MS3

						4	5	6
						13	14	15
						22	23	24
			1	2	3	34	35	36
			10	11	12	43	44	45
			19	20	21	52	53	54
7	8	9	31	32	33	55	56	57
16	17	18	40	41	42	64	65	66
25	26	27	49	50	51	73	74	75
28	29	30	61	62	63			
37	38	39	70	71	72			
46	47	48	79	80	81			
58	59	60						
67	68	69						
76	77	78						

Step 6

The balance of MS3. However, the cell MS3 are still not balance or solve. This we have to solve it one by one.

58	59	60	1	2	3	34	35	36
67	68	69	10	11	12	43	44	45
76	77	78	19	20	21	52	53	54
7	8	9	31	32	33	55	56	57
16	17	18	40	41	42	64	65	66
25	26	27	49	50	51	73	74	75
28	29	30	61	62	63	4	5	6
37	38	39	70	71	72	13	14	15
46	47	48	79	80	81	22	23	24

Step 7

One of the MS3 cell as highlighted in green is solved. See the value as above

77	58	69	1	2	3	34	35	36
60	68	76	10	11	12	43	44	45
67	78	59	19	20	21	52	53	54
7	8	9	31	32	33	55	56	57
16	17	18	40	41	42	64	65	66
25	26	27	49	50	51	73	74	75
28	29	30	61	62	63	4	5	6
37	38	39	70	71	72	13	14	15
46	47	48	79	80	81	22	23	24

			204
77	58	69	204
60	68	76	204
67	78	59	204
204	204	204	204

Step 8

Another one of the MS3 cell as highlighted is solved. See the value as above

77	58	69	20	1	12	34	35	36
60	68	76	3	11	19	43	44	45
67	78	59	10	21	2	52	53	54
7	8	9	31	32	33	55	56	57
16	17	18	40	41	42	64	65	66
25	26	27	49	50	51	73	74	75
28	29	30	61	62	63	4	5	6
37	38	39	70	71	72	13	14	15
46	47	48	79	80	81	22	23	24

			33
20	1	12	33
3	11	19	33
10	21	2	33
33	33	33	33

Step 9

Another one of the MS3 cell as highlighted is solved. See the value as above

77	58	69	20	1	12	53	34	45
60	68	76	3	11	19	36	44	52
67	78	59	10	21	2	43	54	35
7	8	9	31	32	33	55	56	57
16	17	18	40	41	42	64	65	66
25	26	27	49	50	51	73	74	75
28	29	30	61	62	63	4	5	6
37	38	39	70	71	72	13	14	15
46	47	48	79	80	81	22	23	24

			132
53	34	45	132
36	44	52	132
43	54	35	132
132	132	132	132

Step 10

Repeat Step 6 to 8 for below rows, we obtained the solution as highlighted in colour as shown below.

77	58	69	20	1	12	53	34	45
60	68	76	3	11	19	36	44	52
67	78	59	10	21	2	43	54	35
26	7	18	50	31	42	74	55	66
9	17	25	33	41	49	57	65	73
16	27	8	40	51	32	64	75	56
47	28	39	80	61	72	23	4	15
30	38	46	63	71	79	6	14	22
37	48	29	70	81	62	13	24	5

			51
26	7	18	51
9	17	25	51
16	27	8	51
51	51	51	51

			123
50	31	42	123
33	41	49	123
40	51	32	123
123	123	123	123

			195
74	55	66	195
57	65	73	195
64	75	56	195
195	195	195	195

	114		
47	28	39	114
30	38	46	114
37	48	29	114
114	114	114	114

	213		
80	61	72	213
63	71	79	213
70	81	62	213
213	213	213	213

	42		
23	4	15	42
6	14	22	42
13	24	5	42
42	42	42	42

Solution for MS9 (compounded) = MS3 x MS3

77	58	69	20	1	12	53	34	45	369
60	68	76	3	11	19	36	44	52	369
67	78	59	10	21	2	43	54	35	369
26	7	18	50	31	42	74	55	66	369
9	17	25	33	41	49	57	65	73	369
16	27	8	40	51	32	64	75	56	369
47	28	39	80	61	72	23	4	15	369
30	38	46	63	71	79	6	14	22	369
37	48	29	70	81	62	13	24	5	369
369	369	369	369	369	369	369	369	369	369

5.2   Principle of Compounded Magic square?

To understand the principle of compounded magic square let us take the MS 9 for illustration.

Step 1

Starting arrangement

1	2	3	4	5	6	7	8	9
10	11	12	13	14	15	16	17	18
19	20	21	22	23	24	25	26	27
28	29	30	31	32	33	34	35	36
37	38	39	40	41	42	43	44	45
46	47	48	49	50	51	52	53	54
55	56	57	58	59	60	61	62	63
64	65	66	67	68	69	70	71	72
73	74	75	76	77	78	79	80	81

Step 2

Transforming into MS3

1	2	3	4	5	6	7	8	9
10	11	12	13	14	15	16	17	18
19	20	21	22	23	24	25	26	27
28	29	30	31	32	33	34	35	36
37	38	39	40	41	42	43	44	45
46	47	48	49	50	51	52	53	54
55	56	57	58	59	60	61	62	63
64	65	66	67	68	69	70	71	72
73	74	75	76	77	78	79	80	81

Step 3

Summing up all the 9 cells value within the larger MS3. Eg.
(1+2+3+10+11+12+19+20+21) = 99

99	126	153
342	369	396
585	612	639

Step 4

Solving MS3, rotate row clockwise.

	99	126	153
342	369	396	
585	612	639	

Step 5
Putting back to the original square.

153	99	126
342	369	396
612	639	585

Step 6
Rotate column anticlockwise.

		126
	99	396
153	369	585
342	639	
612		

Step 7

Notice that the magic square solved has a value of 1107 or 369 × 3.

612	99	396
153	369	585
342	639	126

			1107
612	99	396	1107
153	369	585	1107
342	639	126	1107
1107	1107	1107	1107

In step 2, we transformed MS9 to MS3 of larger cell. In step 3, take the sum as a whole and form the new MS3. After following the same methodology for MS3, we have a solution for MS in Step 7. The value is 3 times the value of MS9. This means that the transformation formed earlier in step 2 can create a balance force within the larger magic square. Therefore, after going to the principle of yin-yang the larger magic square is solved. However, each MS3 cells has its own cells, altogether 9 cells each. With the similar principle, we performed the balancing for each MS3 cells and thus forming a balance force to the cell level, thus the MS9 comes into existence.

The methodology works because in solving the MS9, we indirectly separate the magic square into movable part (larger MS3) and the immovable part (smaller MS3) in the first stage of solving it. And in the second stage we separate the magic square into the movable part (smaller MS3) and immovable part (larger MS3). Thus, it is just yin-yang principle at play.

5.3 Solving magic square 12 = 4 x 3

Step 1
Starting arrangement.

1	2	3	4	5	6	7	8	9	10	11	12
13	14	15	16	17	18	19	20	21	22	23	24
25	26	27	28	29	30	31	32	33	34	35	36
37	38	39	40	41	42	43	44	45	46	47	48
49	50	51	52	53	54	55	56	57	58	59	60
61	62	63	64	65	66	67	68	69	70	71	72
73	74	75	76	77	78	79	80	81	82	83	84
85	86	87	88	89	90	91	92	93	94	95	96
97	98	99	100	101	102	103	104	105	106	107	108
109	110	111	112	113	114	115	116	117	118	119	120
121	122	123	124	125	126	127	128	129	130	131	132
133	134	135	136	137	138	139	140	141	142	143	144

Step 2

Transforming into MS3 with cells of MS4.

1	2	3	4	5	6	7	8	9	10	11	12
13	14	15	16	17	18	19	20	21	22	23	24
25	26	27	28	29	30	31	32	33	34	35	36
37	38	39	40	41	42	43	44	45	46	47	48
49	50	51	52	53	54	55	56	57	58	59	60
61	62	63	64	65	66	67	68	69	70	71	72
73	74	75	76	77	78	79	80	81	82	83	84
85	86	87	88	89	90	91	92	93	94	95	96
97	98	99	100	101	102	103	104	105	106	107	108
109	110	111	112	113	114	115	116	117	118	119	120
121	122	123	124	125	126	127	128	129	130	131	132
133	134	135	136	137	138	139	140	141	142	143	144

Step 3

Solving larger MS3 on symmetry row, right hand first.

1	2	3	4	5	6	7	8	9	10	11	12
13	14	15	16	17	18	19	20	21	22	23	24
25	26	27	28	29	30	31	32	33	34	35	36
37	38	39	40	41	42	43	44	45	46	47	48
49	50	51	52	53	54	55	56	57	58	59	60
61	62	63	64	65	66	67	68	69	70	71	72
73	74	75	76	77	78	79	80	81	82	83	84
85	86	87	88	89	90	91	92	93	94	95	96
97	98	99	100	101	102	103	104	105	106	107	108
109	110	111	112	113	114	115	116	117	118	119	120
121	122	123	124	125	126	127	128	129	130	131	132
133	134	135	136	137	138	139	140	141	142	143	144

Step 4

Balancing.

9	10	11	12	1	2	3	4	5	6	7	8
21	22	23	24	13	14	15	16	17	18	19	20
33	34	35	36	25	26	27	28	29	30	31	32
45	46	47	48	37	38	39	40	41	42	43	44
49	50	51	52	53	54	55	56	57	58	59	60
61	62	63	64	65	66	67	68	69	70	71	72
73	74	75	76	77	78	79	80	81	82	83	84
85	86	87	88	89	90	91	92	93	94	95	96
97	98	99	100	101	102	103	104	105	106	107	108
109	110	111	112	113	114	115	116	117	118	119	120
121	122	123	124	125	126	127	128	129	130	131	132
133	134	135	136	137	138	139	140	141	142	143	144

Step 5

And then left hand.

9	10	11	12	1	2	3	4	5	6	7	8
21	22	23	24	13	14	15	16	17	18	19	20
33	34	35	36	25	26	27	28	29	30	31	32
45	46	47	48	37	38	39	40	41	42	43	44
49	50	51	52	53	54	55	56	57	58	59	60
61	62	63	64	65	66	67	68	69	70	71	72
73	74	75	76	77	78	79	80	81	82	83	84
85	86	87	88	89	90	91	92	93	94	95	96

97	98	99	100	101	102	103	104	105	106	107	108
109	110	111	112	113	114	115	116	117	118	119	120
121	122	123	124	125	126	127	128	129	130	131	132
133	134	135	136	137	138	139	140	141	142	143	144

Step 6

Balancing.

9	10	11	12	1	2	3	4	5	6	7	8
21	22	23	24	13	14	15	16	17	18	19	20
33	34	35	36	25	26	27	28	29	30	31	32
45	46	47	48	37	38	39	40	41	42	43	44
49	50	51	52	53	54	55	56	57	58	59	60
61	62	63	64	65	66	67	68	69	70	71	72
73	74	75	76	77	78	79	80	81	82	83	84
85	86	87	88	89	90	91	92	93	94	95	96
101	102	103	104	105	106	107	108	97	98	99	100
113	114	115	116	117	118	119	120	109	110	111	112
125	126	127	128	129	130	131	132	121	122	123	124
137	138	139	140	141	142	143	144	133	134	135	136

Step 7

Solving symmetry column.

																5	6	7	8
																17	18	19	20
																29	30	31	32
																41	42	43	44
												1	2	3	4	57	58	59	60
												13	14	15	16	69	70	71	72
												25	26	27	28	81	82	83	84
												37	38	39	40	93	94	95	96
								9	10	11	12	53	54	55	56	97	98	99	100
								21	22	23	24	65	66	67	68	109	110	111	112
								33	34	35	36	77	78	79	80	121	122	123	124
								45	46	47	48	89	90	91	92	133	134	135	136
				49	50	51	52	105	106	107	108								
				61	62	63	64	117	118	119	120								
				73	74	75	76	129	130	131	132								
				85	86	87	88	141	142	143	144								
101	102	103	104																
113	114	115	116																
125	126	127	128																
137	138	139	140																

Step 8

Balancing. And MS3 is solved.

101	102	103	104	1	2	3	4	57	58	59	60
113	114	115	116	13	14	15	16	69	70	71	72
125	126	127	128	25	26	27	28	81	82	83	84
137	138	139	140	37	38	39	40	93	94	95	96
9	10	11	12	53	54	55	56	97	98	99	100
21	22	23	24	65	66	67	68	109	110	111	112
33	34	35	36	77	78	79	80	121	122	123	124
45	46	47	48	89	90	91	92	133	134	135	136
49	50	51	52	105	106	107	108	5	6	7	8
61	62	63	64	117	118	119	120	17	18	19	20
73	74	75	76	129	130	131	132	29	30	31	32
85	86	87	88	141	142	143	144	41	42	43	44

Step 9

Solving the green colour MS4 from Step 8 as shown below.

101	138	139	104	1	2	3	4	57	58	59	60
116	127	126	113	13	14	15	16	69	70	71	72
128	115	114	125	25	26	27	28	81	82	83	84
137	102	103	140	37	38	39	40	93	94	95	96
9	10	11	12	53	54	55	56	97	98	99	100
21	22	23	24	65	66	67	68	109	110	111	112
33	34	35	36	77	78	79	80	121	122	123	124
45	46	47	48	89	90	91	92	133	134	135	136
49	50	51	52	105	106	107	108	5	6	7	8
61	62	63	64	117	118	119	120	17	18	19	20
73	74	75	76	129	130	131	132	29	30	31	32
85	86	87	88	141	142	143	144	41	42	43	44

			482
101	138	139	104	482
116	127	126	113	482
128	115	114	125	482
137	102	103	140	482

482 482 482 482 482

Step 10

And solving another one MS4.

101	138	139	104	1	38	39	4	57	58	59	60
116	127	126	113	16	27	26	13	69	70	71	72
128	115	114	125	28	15	14	25	81	82	83	84
137	102	103	140	37	2	3	40	93	94	95	96
9	10	11	12	53	54	55	56	97	98	99	100
21	22	23	24	65	66	67	68	109	110	111	112
33	34	35	36	77	78	79	80	121	122	123	124
45	46	47	48	89	90	91	92	133	134	135	136
49	50	51	52	105	106	107	108	5	6	7	8
61	62	63	64	117	118	119	120	17	18	19	20
73	74	75	76	129	130	131	132	29	30	31	32
85	86	87	88	141	142	143	144	41	42	43	44

			82
1	38	39	4	82
16	27	26	13	82
28	15	14	25	82
37	2	3	40	82

82 82 82 82 82

Step 11

And another one MS4 solved.

101	138	139	104	1	38	39	4	57	94	95	60
116	127	126	113	16	27	26	13	72	83	82	69
128	115	114	125	28	15	14	25	84	71	70	81
137	102	103	140	37	2	3	40	93	58	59	96
9	10	11	12	53	54	55	56	97	98	99	100
21	22	23	24	65	66	67	68	109	110	111	112
33	34	35	36	77	78	79	80	121	122	123	124
45	46	47	48	89	90	91	92	133	134	135	136
49	50	51	52	105	106	107	108	5	6	7	8
61	62	63	64	117	118	119	120	17	18	19	20
73	74	75	76	129	130	131	132	29	30	31	32
85	86	87	88	141	142	143	144	41	42	43	44

				306
57	94	95	60	306
72	83	82	69	306
84	71	70	81	306
93	58	59	96	306
306	306	306	306	306

Step 12

And solving all the MS4 as shown below.

101	138	139	104	1	38	39	4	57	94	95	60
116	127	126	113	16	27	26	13	72	83	82	69
128	115	114	125	28	15	14	25	84	71	70	81
137	102	103	140	37	2	3	40	93	58	59	96
9	46	47	12	53	90	91	56	97	134	135	100
24	35	34	21	68	79	78	65	112	123	122	109
36	23	22	33	80	67	66	77	124	111	110	121
45	10	11	48	89	54	55	92	133	98	99	136
49	86	87	52	105	142	143	108	5	42	43	8
64	75	74	61	120	131	130	117	20	31	30	17
76	63	62	73	132	119	118	129	32	19	18	29
85	50	51	88	141	106	107	144	41	6	7	44

114

9	46	47	12	114
24	35	34	21	114
36	23	22	33	114
45	10	11	48	114
114	114	114	114	114

290

53	90	91	56	290
68	79	78	65	290
80	67	66	77	290
89	54	55	92	290
290	290	290	290	290

466

97	134	135	100	466
112	123	122	109	466
124	111	110	121	466
133	98	99	136	466
466	466	466	466	466

274

49	86	87	52	274
64	75	74	61	274
76	63	62	73	274
85	50	51	88	274
274	274	274	274	274

498

105	142	143	108	498
120	131	130	117	498
132	119	118	129	498
141	106	107	144	498
498	498	498	498	498

98

5	42	43	8	98
20	31	30	17	98
32	19	18	29	98
41	6	7	44	98
98	98	98	98	98

Solution of MS12 in 4x3

870

101	138	139	104	1	38	39	4	57	94	95	60	870
116	127	126	113	16	27	26	13	72	83	82	69	870
128	115	114	125	28	15	14	25	84	71	70	81	870
137	102	103	140	37	2	3	40	93	58	59	96	870
9	46	47	12	53	90	91	56	97	134	135	100	870
24	35	34	21	68	79	78	65	112	123	122	109	870
36	23	22	33	80	67	66	77	124	111	110	121	870
45	10	11	48	89	54	55	92	133	98	99	136	870
49	86	87	52	105	142	143	108	5	42	43	8	870
64	75	74	61	120	131	130	117	20	31	30	17	870
76	63	62	73	132	119	118	129	32	19	18	29	870
85	50	51	88	141	106	107	144	41	6	7	44	870
870	870	870	870	870	870	870	870	870	870	870	870	870

Question 12

How many solution of MS12 can you obtain using this method? Thousand? Millions? Billions? It should be more than that.

The methodology of sovling MS12 above is based on the transformation 12 = 4 x 3. This transformation stated that the MS12 is converted into MS3 but with cells MS4. The solving can be done on the smaller magic square then the larger. On the other hand, if the transformation is based on 12 = 3 x 4, then we will have a totally new solution to MS12. Let us see below.

Magic Square 12 = 4 x 3

Step 1

Starting arrangement.

1	2	3	4	5	6	7	8	9	10	11	12
13	14	15	16	17	18	19	20	21	22	23	24
25	26	27	28	29	30	31	32	33	34	35	36
37	38	39	40	41	42	43	44	45	46	47	48
49	50	51	52	53	54	55	56	57	58	59	60
61	62	63	64	65	66	67	68	69	70	71	72
73	74	75	76	77	78	79	80	81	82	83	84
85	86	87	88	89	90	91	92	93	94	95	96
97	98	99	100	101	102	103	104	105	106	107	108
109	110	111	112	113	114	115	116	117	118	119	120
121	122	123	124	125	126	127	128	129	130	131	132
133	134	135	136	137	138	139	140	141	142	143	144

Step 2

Transforming into MS4 with cells of MS3.

1	2	3	4	5	6	7	8	9	10	11	12
13	14	15	16	17	18	19	20	21	22	23	24
25	26	27	28	29	30	31	32	33	34	35	36
37	38	39	40	41	42	43	44	45	46	47	48
49	50	51	52	53	54	55	56	57	58	59	60
61	62	63	64	65	66	67	68	69	70	71	72
73	74	75	76	77	78	79	80	81	82	83	84
85	86	87	88	89	90	91	92	93	94	95	96
97	98	99	100	101	102	103	104	105	106	107	108
109	110	111	112	113	114	115	116	117	118	119	120
121	122	123	124	125	126	127	128	129	130	131	132
133	134	135	136	137	138	139	140	141	142	143	144

Step 3

Rotation 180^0 about the horizontal axis.

1	2	3	112	113	114	115	116	117	10	11	12
13	14	15	124	125	126	127	128	129	22	23	24
25	26	27	136	137	138	139	140	141	34	35	36
37	38	39	76	77	78	79	80	81	46	47	48
49	50	51	88	89	90	91	92	93	58	59	60
61	62	63	100	101	102	103	104	105	70	71	72
73	74	75	40	41	42	43	44	45	82	83	84
85	86	87	52	53	54	55	56	57	94	95	96
97	98	99	64	65	66	67	68	69	106	107	108
109	110	111	4	5	6	7	8	9	118	119	120
121	122	123	16	17	18	19	20	21	130	131	132
133	134	135	28	29	30	31	32	33	142	143	144

Step 4

Rotation 180⁰ about the vertical axis. MS4 is solved.

1	2	3	112	113	114	115	116	117	10	11	12
13	14	15	124	125	126	127	128	129	22	23	24
25	26	27	136	137	138	139	140	141	34	35	36
46	47	48	79	80	81	76	77	78	37	38	39
58	59	60	91	92	93	88	89	90	49	50	51
70	71	72	103	104	105	100	101	102	61	62	63
82	83	84	43	44	45	40	41	42	73	74	75
94	95	96	55	56	57	52	53	54	85	86	87
106	107	108	67	68	69	64	65	66	97	98	99
109	110	111	4	5	6	7	8	9	118	119	120
121	122	123	16	17	18	19	20	21	130	131	132
133	134	135	28	29	30	31	32	33	142	143	144

Step 5

Solving one of the MS3 from Step 4.

26	1	15	112	113	114	115	116	117	10	11	12
3	14	25	124	125	126	127	128	129	22	23	24
13	27	2	136	137	138	139	140	141	34	35	36
46	47	48	79	80	81	76	77	78	37	38	39
58	59	60	91	92	93	88	89	90	49	50	51
70	71	72	103	104	105	100	101	102	61	62	63
82	83	84	43	44	45	40	41	42	73	74	75
94	95	96	55	56	57	52	53	54	85	86	87
106	107	108	67	68	69	64	65	66	97	98	99
109	110	111	4	5	6	7	8	9	118	119	120
121	122	123	16	17	18	19	20	21	130	131	132
133	134	135	28	29	30	31	32	33	142	143	144

			42
26	1	15	42
3	14	25	42
13	27	2	42
42	42	42	42

Step 6

And another MS3 is solved.

26	1	15	137	112	126	115	116	117	10	11	12
3	14	25	114	125	136	127	128	129	22	23	24
13	27	2	124	138	113	139	140	141	34	35	36
46	47	48	79	80	81	76	77	78	37	38	39
58	59	60	91	92	93	88	89	90	49	50	51
70	71	72	103	104	105	100	101	102	61	62	63
82	83	84	43	44	45	40	41	42	73	74	75
94	95	96	55	56	57	52	53	54	85	86	87
106	107	108	67	68	69	64	65	66	97	98	99
109	110	111	4	5	6	7	8	9	118	119	120
121	122	123	16	17	18	19	20	21	130	131	132
133	134	135	28	29	30	31	32	33	142	143	144

375

137	112	126
114	125	136
124	138	113

375
375
375
375 375 375 375

Step 7

Repeat Step 5 or 6 for the rest of the MS3, thus the solution is as below.

26	1	15	137	112	126	140	115	129	35	10	24
3	14	25	114	125	136	117	128	139	12	23	34
13	27	2	124	138	113	127	141	116	22	36	11
71	46	60	104	79	93	101	76	90	62	37	51
48	59	70	81	92	103	78	89	100	39	50	61
58	72	47	91	105	80	88	102	77	49	63	38
107	82	96	68	43	57	65	40	54	98	73	87
84	95	106	45	56	67	42	53	64	75	86	97
94	108	83	55	69	44	52	66	41	85	99	74
134	109	123	29	4	18	32	7	21	143	118	132
111	122	133	6	17	28	9	20	31	120	131	142
121	135	110	16	30	5	19	33	8	130	144	119

Solution for MS12 = 3x4

												870
26	1	15	137	112	126	140	115	129	35	10	24	870
3	14	25	114	125	136	117	128	139	12	23	34	870
13	27	2	124	138	113	127	141	116	22	36	11	870
71	46	60	104	79	93	101	76	90	62	37	51	870
48	59	70	81	92	103	78	89	100	39	50	61	870
58	72	47	91	105	80	88	102	77	49	63	38	870
107	82	96	68	43	57	65	40	54	98	73	87	870
84	95	106	45	56	67	42	53	64	75	86	97	870
94	108	83	55	69	44	52	66	41	85	99	74	870
134	109	123	29	4	18	32	7	21	143	118	132	870
111	122	133	6	17	28	9	20	31	120	131	142	870
121	135	110	16	30	5	19	33	8	130	144	119	870

870 870 870 870 870 870 870 870 870 870 870 870 870

Question 13

How many solution of MS12 can you obtain using this method? Thousand? Millions? Billions? It should be more than that.

Thus, to summarised the methodology, we can solve magic square in many ways and obtain the variation in many ways. Table 4 below shows the method of solving magic square from 3 to 30.

MS	Methodology				
3	2N + 1				
4	4N				
5	2N + 1				
6	4N + 2				

7	2N + 1				
8	4N				
9	2N + 1	3 x 3			
10	4N + 2				
11	2N + 1				
12	4N	4 x 3	3 x 4		
13	2N + 1				
14	4N + 2				
15	2N + 1	3 x 5	5 x 3		
16	4N	4 x 4			
17	2N + 1				
18	4N + 2	3 X 6	6 X 3		
19	2N + 1				
20	4N	4 X 5	5 X 4		
21	2N + 1	3 X 7	7 X 3		
22	4N + 2				
23	2N + 1				
24	4N	4 X 6	6 x4	3 x 8	8 x 3
25	2N + 1	5 X 5			
26	4N + 2				
27	2N + 1	3 X 9	9 X 3	3 x 3 x 3	
28	4N	4 X 7	7 X 4		
29	2N + 1				
30	4N + 2	5 X 6	6 X 5		

From the principle above, as long as magic square N (MSN) can be formed into a factor, then we can solve magic square by this compounded method. Let say we have.

$$N = a \times b \times c$$
$$N = a \times b \times c \times d$$
$$N = a \times b \times c \times d \times e$$

We can solve magic square a, b, c, d and e one by one separately. For the case of $N = a \times b \times c$, the smallest unit is $3 \times 3 \times 3 = 27$. See Appendix 2 for the solution.

Question 14

What is the smallest unit for case $N = a \times b \times c \times d$? And for case $N = a \times b \times c \times d \times e$?

Magic Square 18, (4N + 2, where N = 4)

(Solution to MS18 solving by putting N = 4)

201	202	241	242	281	282	321	304	20	19	42	41	82	81	122	121	162	161
183	184	223	224	263	264	303	322	2	1	60	59	100	99	140	139	180	179
239	240	279	280	319	320	35	18	58	57	80	79	120	119	160	159	182	181
221	222	261	262	301	302	17	36	40	39	98	97	138	137	178	177	200	199
277	278	317	318	33	34	55	38	96	95	118	117	158	157	198	197	220	219
259	260	299	300	15	16	37	56	78	77	136	135	176	175	216	215	238	237
315	316	31	32	71	72	93	76	134	133	156	155	196	195	218	217	258	257
297	298	13	14	53	54	75	94	115	116	173	174	213	214	235	236	275	276
12	11	52	51	74	73	114	113	172	171	194	211	252	251	274	273	314	313
30	29	70	69	92	91	132	131	154	153	212	193	234	233	256	255	296	295
49	50	89	90	111	112	151	152	209	210	231	250	271	272	311	312	27	28
68	67	108	107	130	129	170	169	191	192	249	232	253	254	293	294	9	10
88	87	110	109	150	149	190	189	247	248	269	288	309	310	25	26	65	66
106	105	128	127	168	167	208	207	229	230	287	270	291	292	7	8	47	48
126	125	148	147	188	187	228	227	285	286	289	308	23	24	63	64	103	104
144	143	166	165	206	205	246	245	267	268	307	290	5	6	45	46	85	86
146	145	186	185	226	225	266	265	323	324	3	22	61	62	101	102	141	142
164	163	204	203	244	243	284	283	305	306	21	4	43	44	83	84	123	124

Can you solve this?

Hint:
    a) Transform the magic square as **MS18 = MS2 x MS9**.
    b) Solve the **MS9** as odd magic square.
    c) Interchange the cells within the **MS2** of the symmetry column.
    d) Interchange the cells within the **MS2** of the symmetry row.
    e) Interchange the cells within the **MS2** following the pattern. (Note **N** = 4)

Magic Square 18 = 6 x 3 (compounded)

(Solution to MS18 = 6X3 method).

315	298	242	241	264	263	93	76	20	19	42	41	213	196	140	139	162	161
297	316	223	224	281	282	75	94	1	2	59	60	195	214	121	122	179	180
228	227	280	279	296	313	6	5	58	57	74	91	126	125	178	177	194	211
246	245	262	261	314	295	24	23	40	39	92	73	144	143	160	159	212	193
259	260	317	318	225	244	37	38	95	96	3	22	157	158	215	216	123	142
278	277	299	300	243	226	56	55	77	78	21	4	176	175	197	198	141	124
105	88	32	31	54	53	207	190	134	133	156	155	309	292	236	235	258	257
87	106	13	14	71	72	189	208	115	116	173	174	291	310	217	218	275	276
18	17	70	69	86	103	120	119	172	171	188	205	222	221	274	273	290	307
36	35	52	51	104	85	138	137	154	153	206	187	240	239	256	255	308	289
49	50	107	108	15	34	151	152	209	210	117	136	253	254	311	312	219	238
68	67	89	90	33	16	170	169	191	192	135	118	272	271	293	294	237	220
201	184	128	127	150	149	321	304	248	247	270	269	99	82	26	25	48	47
183	202	109	110	167	168	303	322	229	230	287	288	81	100	7	8	65	66
114	113	166	165	182	199	234	233	286	285	302	319	12	11	64	63	80	97
132	131	148	147	200	181	252	251	268	267	320	301	30	29	46	45	98	79
145	146	203	204	111	130	265	266	323	324	231	250	43	44	101	102	9	28
164	163	185	186	129	112	284	283	305	306	249	232	62	61	83	84	27	10

Can you solve this?

Hint:
- a) Transform the magic square as **MS18 = MS6 x MS3**.
- b) Solve the smaller **MS6** one by one.
- c) Solve the larger **MS3**.

Magic Square 18 = 3 x 6 (compounded)

(Solution to MS18 = 3 x 6 method).

314	277	297	263	226	246	95	58	78	92	55	75	161	124	144	158	121	141
279	296	313	228	245	262	60	77	94	57	74	91	126	143	160	123	140	157
295	315	278	244	264	227	76	96	59	73	93	56	142	162	125	139	159	122
260	223	243	317	280	300	38	1	21	41	4	24	212	175	195	215	178	198
225	242	259	282	299	316	3	20	37	6	23	40	177	194	211	180	197	214
241	261	224	298	318	281	19	39	2	22	42	5	193	213	176	196	216	179
53	16	36	50	13	33	209	172	192	206	169	189	257	220	240	308	271	291
18	35	52	15	32	49	174	191	208	171	188	205	222	239	256	273	290	307
34	54	17	31	51	14	190	210	173	187	207	170	238	258	221	289	309	272
107	70	90	104	67	87	155	118	138	152	115	135	311	274	294	254	217	237
72	89	106	69	86	103	120	137	154	117	134	151	276	293	310	219	236	253
88	108	71	85	105	68	136	156	119	133	153	116	292	312	275	235	255	218
146	109	129	149	112	132	320	283	303	323	286	306	44	7	27	101	64	84
111	128	145	114	131	148	285	302	319	288	305	322	9	26	43	66	83	100
127	147	110	130	150	113	301	321	284	304	324	287	25	45	8	82	102	65
203	166	186	200	163	183	266	229	249	269	232	252	98	61	81	47	10	30
168	185	202	165	182	199	231	248	265	234	251	268	63	80	97	12	29	46
184	204	167	181	201	164	247	267	230	250	270	233	79	99	62	28	48	11

Can you solve this?

Hint:

    a) Transform the magic square as **MS18 = MS3 x MS6**.

    b) Solve the smaller **MS3** one by one.

    c) Solve the larger **MS6**.

Magic Square 3x3x3

Step 1

Starting arrangement.

1	2	3	4	5	6	7	8	9	10	11	12	13	14	15	16	17	18	19	20	21	22	23	24	25	26	27
28	29	30	31	32	33	34	35	36	37	38	39	40	41	42	43	44	45	46	47	48	49	50	51	52	53	54
55	56	57	58	59	60	61	62	63	64	65	66	67	68	69	70	71	72	73	74	75	76	77	78	79	80	81
82	83	84	85	86	87	88	89	90	91	92	93	94	95	96	97	98	99	100	101	102	103	104	105	106	107	108
109	110	111	112	113	114	115	116	117	118	119	120	121	122	123	124	125	126	127	128	129	130	131	132	133	134	135
136	137	138	139	140	141	142	143	144	145	146	147	148	149	150	151	152	153	154	155	156	157	158	159	160	161	162
163	164	165	166	167	168	169	170	171	172	173	174	175	176	177	178	179	180	181	182	183	184	185	186	187	188	189
190	191	192	193	194	195	196	197	198	199	200	201	202	203	204	205	206	207	208	209	210	211	212	213	214	215	216
217	218	219	220	221	222	223	224	225	226	227	228	229	230	231	232	233	234	235	236	237	238	239	240	241	242	243
244	245	246	247	248	249	250	251	252	253	254	255	256	257	258	259	260	261	262	263	264	265	266	267	268	269	270
271	272	273	274	275	276	277	278	279	280	281	282	283	284	285	286	287	288	289	290	291	292	293	294	295	296	297
298	299	300	301	302	303	304	305	306	307	308	309	310	311	312	313	314	315	316	317	318	319	320	321	322	323	324
325	326	327	328	329	330	331	332	333	334	335	336	337	338	339	340	341	342	343	344	345	346	347	348	349	350	351
352	353	354	355	356	357	358	359	360	361	362	363	364	365	366	367	368	369	370	371	372	373	374	375	376	377	378
379	380	381	382	383	384	385	386	387	388	389	390	391	392	393	394	395	396	397	398	399	400	401	402	403	404	405
406	407	408	409	410	411	412	413	414	415	416	417	418	419	420	421	422	423	424	425	426	427	428	429	430	431	432
433	434	435	436	437	438	439	440	441	442	443	444	445	446	447	448	449	450	451	452	453	454	455	456	457	458	459
460	461	462	463	464	465	466	467	468	469	470	471	472	473	474	475	476	477	478	479	480	481	482	483	484	485	486
487	488	489	490	491	492	493	494	495	496	497	498	499	500	501	502	503	504	505	506	507	508	509	510	511	512	513
514	515	516	517	518	519	520	521	522	523	524	525	526	527	528	529	530	531	532	533	534	535	536	537	538	539	540
541	542	543	544	545	546	547	548	549	550	551	552	553	554	555	556	557	558	559	560	561	562	563	564	565	566	567
568	569	570	571	572	573	574	575	576	577	578	579	580	581	582	583	584	585	586	587	588	589	590	591	592	593	594
595	596	597	598	599	600	601	602	603	604	605	606	607	608	609	610	611	612	613	614	615	616	617	618	619	620	621
622	623	624	625	626	627	628	629	630	631	632	633	634	635	636	637	638	639	640	641	642	643	644	645	646	647	648
649	650	651	652	653	654	655	656	657	658	659	660	661	662	663	664	665	666	667	668	669	670	671	672	673	674	675
676	677	678	679	680	681	682	683	684	685	686	687	688	689	690	691	692	693	694	695	696	697	698	699	700	701	702
703	704	705	706	707	708	709	710	711	712	713	714	715	716	717	718	719	720	721	722	723	724	725	726	727	728	729

Step 2

Transform 27 = 9x3 or MS27 = 9xMS3 as shown by the bold line.

1	2	3	4	5	6	7	8	9	10	11	12	13	14	15	16	17	18	19	20	21	22	23	24	25	26	27
28	29	30	31	32	33	34	35	36	37	38	39	40	41	42	43	44	45	46	47	48	49	50	51	52	53	54
55	56	57	58	59	60	61	62	63	64	65	66	67	68	69	70	71	72	73	74	75	76	77	78	79	80	81
82	83	84	85	86	87	88	89	90	91	92	93	94	95	96	97	98	99	100	101	102	103	104	105	106	107	108
109	110	111	112	113	114	115	116	117	118	119	120	121	122	123	124	125	126	127	128	129	130	131	132	133	134	135
136	137	138	139	140	141	142	143	144	145	146	147	148	149	150	151	152	153	154	155	156	157	158	159	160	161	162
163	164	165	166	167	168	169	170	171	172	173	174	175	176	177	178	179	180	181	182	183	184	185	186	187	188	189
190	191	192	193	194	195	196	197	198	199	200	201	202	203	204	205	206	207	208	209	210	211	212	213	214	215	216
217	218	219	220	221	222	223	224	225	226	227	228	229	230	231	232	233	234	235	236	237	238	239	240	241	242	243
244	245	246	247	248	249	250	251	252	253	254	255	256	257	258	259	260	261	262	263	264	265	266	267	268	269	270
271	272	273	274	275	276	277	278	279	280	281	282	283	284	285	286	287	288	289	290	291	292	293	294	295	296	297
298	299	300	301	302	303	304	305	306	307	308	309	310	311	312	313	314	315	316	317	318	319	320	321	322	323	324
325	326	327	328	329	330	331	332	333	334	335	336	337	338	339	340	341	342	343	344	345	346	347	348	349	350	351
352	353	354	355	356	357	358	359	360	361	362	363	364	365	366	367	368	369	370	371	372	373	374	375	376	377	378
379	380	381	382	383	384	385	386	387	388	389	390	391	392	393	394	395	396	397	398	399	400	401	402	403	404	405
406	407	408	409	410	411	412	413	414	415	416	417	418	419	420	421	422	423	424	425	426	427	428	429	430	431	432
433	434	435	436	437	438	439	440	441	442	443	444	445	446	447	448	449	450	451	452	453	454	455	456	457	458	459
460	461	462	463	464	465	466	467	468	469	470	471	472	473	474	475	476	477	478	479	480	481	482	483	484	485	486
487	488	489	490	491	492	493	494	495	496	497	498	499	500	501	502	503	504	505	506	507	508	509	510	511	512	513
514	515	516	517	518	519	520	521	522	523	524	525	526	527	528	529	530	531	532	533	534	535	536	537	538	539	540
541	542	543	544	545	546	547	548	549	550	551	552	553	554	555	556	557	558	559	560	561	562	563	564	565	566	567
568	569	570	571	572	573	574	575	576	577	578	579	580	581	582	583	584	585	586	587	588	589	590	591	592	593	594
595	596	597	598	599	600	601	602	603	604	605	606	607	608	609	610	611	612	613	614	615	616	617	618	619	620	621
622	623	624	625	626	627	628	629	630	631	632	633	634	635	636	637	638	639	640	641	642	643	644	645	646	647	648
649	650	651	652	653	654	655	656	657	658	659	660	661	662	663	664	665	666	667	668	669	670	671	672	673	674	675
676	677	678	679	680	681	682	683	684	685	686	687	688	689	690	691	692	693	694	695	696	697	698	699	700	701	702
703	704	705	706	707	708	709	710	711	712	713	714	715	716	717	718	719	720	721	722	723	724	725	726	727	728	729

Step 3

Transform MS27 = 3xMS3xMS3 as shown by the double line.

1	2	3	4	5	6	7	8	9	10	11	12	13	14	15	16	17	18	19	20	21	22	23	24	25	26	27
28	29	30	31	32	33	34	35	36	37	38	39	40	41	42	43	44	45	46	47	48	49	50	51	52	53	54
55	56	57	58	59	60	61	62	63	64	65	66	67	68	69	70	71	72	73	74	75	76	77	78	79	80	81
82	83	84	85	86	87	88	89	90	91	92	93	94	95	96	97	98	99	100	101	102	103	104	105	106	107	108
109	110	111	112	113	114	115	116	117	118	119	120	121	122	123	124	125	126	127	128	129	130	131	132	133	134	135
136	137	138	139	140	141	142	143	144	145	146	147	148	149	150	151	152	153	154	155	156	157	158	159	160	161	162
163	164	165	166	167	168	169	170	171	172	173	174	175	176	177	178	179	180	181	182	183	184	185	186	187	188	189
190	191	192	193	194	195	196	197	198	199	200	201	202	203	204	205	206	207	208	209	210	211	212	213	214	215	216
217	218	219	220	221	222	223	224	225	226	227	228	229	230	231	232	233	234	235	236	237	238	239	240	241	242	243
244	245	246	247	248	249	250	251	252	253	254	255	256	257	258	259	260	261	262	263	264	265	266	267	268	269	270
271	272	273	274	275	276	277	278	279	280	281	282	283	284	285	286	287	288	289	290	291	292	293	294	295	296	297
298	299	300	301	302	303	304	305	306	307	308	309	310	311	312	313	314	315	316	317	318	319	320	321	322	323	324
325	326	327	328	329	330	331	332	333	334	335	336	337	338	339	340	341	342	343	344	345	346	347	348	349	350	351
352	353	354	355	356	357	358	359	360	361	362	363	364	365	366	367	368	369	370	371	372	373	374	375	376	377	378
379	380	381	382	383	384	385	386	387	388	389	390	391	392	393	394	395	396	397	398	399	400	401	402	403	404	405
406	407	408	409	410	411	412	413	414	415	416	417	418	419	420	421	422	423	424	425	426	427	428	429	430	431	432
433	434	435	436	437	438	439	440	441	442	443	444	445	446	447	448	449	450	451	452	453	454	455	456	457	458	459
460	461	462	463	464	465	466	467	468	469	470	471	472	473	474	475	476	477	478	479	480	481	482	483	484	485	486
487	488	489	490	491	492	493	494	495	496	497	498	499	500	501	502	503	504	505	506	507	508	509	510	511	512	513
514	515	516	517	518	519	520	521	522	523	524	525	526	527	528	529	530	531	532	533	534	535	536	537	538	539	540
541	542	543	544	545	546	547	548	549	550	551	552	553	554	555	556	557	558	559	560	561	562	563	564	565	566	567
568	569	570	571	572	573	574	575	576	577	578	579	580	581	582	583	584	585	586	587	588	589	590	591	592	593	594
595	596	597	598	599	600	601	602	603	604	605	606	607	608	609	610	611	612	613	614	615	616	617	618	619	620	621
622	623	624	625	626	627	628	629	630	631	632	633	634	635	636	637	638	639	640	641	642	643	644	645	646	647	648
649	650	651	652	653	654	655	656	657	658	659	660	661	662	663	664	665	666	667	668	669	670	671	672	673	674	675
676	677	678	679	680	681	682	683	684	685	686	687	688	689	690	691	692	693	694	695	696	697	698	699	700	701	702
703	704	705	706	707	708	709	710	711	712	713	714	715	716	717	718	719	720	721	722	723	724	725	726	727	728	729

Step 4

Solve one by one 81 MS3 within double line.

56	1	30	59	4	33	62	7	36	65	10	39	68	13	42	71	16	45	74	19	48	77	22	51	80	25	54
3	29	55	6	32	58	9	35	61	12	38	64	15	41	67	18	44	70	21	47	73	24	50	76	27	53	79
28	57	2	31	60	5	34	63	8	37	66	11	40	69	14	43	72	17	46	75	20	49	78	23	52	81	26
137	82	111	140	85	114	143	88	117	146	91	120	149	94	123	152	97	126	155	100	129	158	103	132	161	106	135
84	110	136	87	113	139	90	116	142	93	119	145	96	122	148	99	125	151	102	128	154	105	131	157	108	134	160
109	138	83	112	141	86	115	144	89	118	147	92	121	150	95	124	153	98	127	156	101	130	159	104	133	162	107
218	163	192	221	166	195	224	169	198	227	172	201	230	175	204	233	178	207	236	181	210	239	184	213	242	187	216
165	191	217	168	194	220	171	197	223	174	200	226	177	203	229	180	206	232	183	209	235	186	212	238	189	215	241
190	219	164	193	222	167	196	225	170	199	228	173	202	231	176	205	234	179	208	237	182	211	240	185	214	243	188
299	244	273	302	247	276	305	250	279	308	253	282	311	256	285	314	259	288	317	262	291	320	265	294	323	268	297
246	272	298	249	275	301	252	278	304	255	281	307	258	284	310	261	287	313	264	290	316	267	293	319	270	296	322
271	300	245	274	303	248	277	306	251	280	309	254	283	312	257	286	315	260	289	318	263	292	321	266	295	324	269
380	325	354	383	328	357	386	331	360	389	334	363	392	337	366	395	340	369	398	343	372	401	346	375	404	349	378
327	353	379	330	356	382	333	359	385	336	362	388	339	365	391	342	368	394	345	371	397	348	374	400	351	377	403
352	381	326	355	384	329	358	387	332	361	390	335	364	393	338	367	396	341	370	399	344	373	402	347	376	405	350
461	406	435	464	409	438	467	412	441	470	415	444	473	418	447	476	421	450	479	424	453	482	427	456	485	430	459
408	434	460	411	437	463	414	440	466	417	443	469	420	446	472	423	449	475	426	452	478	429	455	481	432	458	484
433	462	407	436	465	410	439	468	413	442	471	416	445	474	419	448	477	422	451	480	425	454	483	428	457	486	431
542	487	516	545	490	519	548	493	522	551	496	525	554	499	528	557	502	531	560	505	534	563	508	537	566	511	540
489	515	541	492	518	544	495	521	547	498	524	550	501	527	553	504	530	556	507	533	559	510	536	562	513	539	565
514	543	488	517	546	491	520	549	494	523	552	497	526	555	500	529	558	503	532	561	506	535	564	509	538	567	512
623	568	597	626	571	600	629	574	603	632	577	606	635	580	609	638	583	612	641	586	615	644	589	618	647	592	621
570	596	622	573	599	625	576	602	628	579	605	631	582	608	634	585	611	637	588	614	640	591	617	643	594	620	646
595	624	569	598	627	572	601	630	575	604	633	578	607	636	581	610	639	584	613	642	587	616	645	590	619	648	593
704	649	678	707	652	681	710	655	684	713	658	687	716	661	690	719	664	693	722	667	696	725	670	699	728	673	702
651	677	703	654	680	706	657	683	709	660	686	712	663	689	715	666	692	718	669	695	721	672	698	724	675	701	727
676	705	650	679	708	653	682	711	656	685	714	659	688	717	662	691	720	665	694	723	668	697	726	671	700	729	674

Step 5

Solve one by one 9 MS9 within red bold line

221	166	195	56	1	30	143	88	117	230	175	204	65	10	39	152	97	126	482	427	456	317	262	291	404	349	378
168	194	220	3	29	55	90	116	142	177	203	229	12	38	64	99	125	151	429	455	481	264	290	316	351	377	403
193	222	167	28	57	2	115	144	89	202	231	176	37	66	11	124	153	98	454	483	428	289	318	263	376	405	350
62	7	36	140	85	114	218	163	192	71	16	45	149	94	123	227	172	201	323	268	297	401	346	375	479	424	453
9	35	61	87	113	139	165	191	217	18	44	70	96	122	148	174	200	226	270	296	322	348	374	400	426	452	478
34	63	8	112	141	86	190	219	164	43	72	17	121	150	95	199	228	173	295	324	269	373	402	347	451	480	425
137	82	111	224	169	198	59	4	33	146	91	120	233	178	207	68	13	42	398	343	372	485	430	459	320	265	294
84	110	136	171	197	223	6	32	58	93	119	145	180	206	232	15	41	67	345	371	397	432	458	484	267	293	319
109	138	83	196	225	170	31	60	5	118	147	92	205	234	179	40	69	14	370	399	344	457	486	431	292	321	266
464	409	438	299	244	273	386	331	360	473	418	447	308	253	282	395	340	369	239	184	213	74	19	48	161	106	135
411	437	463	246	272	298	333	359	385	420	446	472	255	281	307	342	368	394	186	212	238	21	47	73	108	134	160
436	465	410	271	300	245	358	387	332	445	474	419	280	309	254	367	396	341	211	240	185	46	75	20	133	162	107
305	250	279	383	328	357	461	406	435	314	259	288	392	337	366	470	415	444	80	25	54	158	103	132	236	181	210
252	278	304	330	356	382	408	434	460	261	287	313	339	365	391	417	443	469	27	53	79	105	131	157	183	209	235
277	306	251	355	384	329	433	462	407	286	315	260	364	393	338	442	471	416	52	81	26	130	159	104	208	237	182
380	325	354	467	412	441	302	247	276	389	334	363	476	421	450	311	256	285	155	100	129	242	187	216	77	22	51
327	353	379	414	440	466	249	275	301	336	362	388	423	449	475	258	284	310	102	128	154	189	215	241	24	50	76
352	381	326	439	468	413	274	303	248	361	390	335	448	477	422	283	312	257	127	156	101	214	243	188	49	78	23
707	652	681	542	487	516	629	574	603	716	661	690	551	496	525	638	583	612	725	670	699	560	505	534	647	592	621
654	680	706	489	515	541	576	602	628	663	689	715	498	524	550	585	611	637	672	698	724	507	533	559	594	620	646
679	708	653	514	543	488	601	630	575	688	717	662	523	552	497	610	639	584	697	726	671	532	561	506	619	648	593
548	493	522	626	571	600	704	649	678	557	502	531	635	580	609	713	658	687	566	511	540	644	589	618	722	667	696
495	521	547	573	599	625	651	677	703	504	530	556	582	608	634	660	686	712	513	539	565	591	617	643	669	695	721
520	549	494	598	627	572	676	705	650	529	558	503	607	636	581	685	714	659	538	567	512	616	645	590	694	723	668
623	568	597	710	655	684	545	490	519	632	577	606	719	664	693	554	499	528	641	586	615	728	673	702	563	508	537
570	596	622	657	683	709	492	518	544	579	605	631	666	692	718	501	527	553	588	614	640	675	701	727	510	536	562
595	624	569	682	711	656	517	546	491	604	633	578	691	720	665	526	555	500	613	642	587	700	729	674	535	564	509

Step 6

Solve MS27 with cells MS9. You have just solved MS27 with the expression 3x3x3. Congratulation

716	661	690	551	496	525	638	583	612	221	166	195	56	1	30	143	88	117	482	427	456	317	262	291	404	349	378
663	689	715	498	524	550	585	611	637	168	194	220	3	29	55	90	116	142	429	455	481	264	290	316	351	377	403
688	717	662	523	552	497	610	639	584	193	222	167	28	57	2	115	144	89	454	483	428	289	318	263	376	405	350
557	502	531	635	580	609	713	658	687	62	7	36	140	85	114	218	163	192	323	268	297	401	346	375	479	424	453
504	530	556	582	608	634	660	686	712	9	35	61	87	113	139	165	191	217	270	296	322	348	374	400	426	452	478
529	558	503	607	636	581	685	714	659	34	63	8	112	141	86	190	219	164	295	324	269	373	402	347	451	480	425
632	577	606	719	664	693	554	499	528	137	82	111	224	169	198	59	4	33	398	343	372	485	430	459	320	265	294
579	605	631	666	692	718	501	527	553	84	110	136	171	197	223	6	32	58	345	371	397	432	458	484	267	293	319
604	633	578	691	720	665	526	555	500	109	138	83	196	225	170	31	60	5	370	399	344	457	486	431	292	321	266
239	184	213	74	19	48	161	106	135	473	418	447	308	253	282	395	340	369	707	652	681	542	487	516	629	574	603
186	212	238	21	47	73	108	134	160	420	446	472	255	281	307	342	368	394	654	680	706	489	515	541	576	602	628
211	240	185	46	75	20	133	162	107	445	474	419	280	309	254	367	396	341	679	708	653	514	543	488	601	630	575
80	25	54	158	103	132	236	181	210	314	259	288	392	337	366	470	415	444	548	493	522	626	571	600	704	649	678
27	53	79	105	131	157	183	209	235	261	287	313	339	365	391	417	443	469	495	521	547	573	599	625	651	677	703
52	81	26	130	159	104	208	237	182	286	315	260	364	393	338	442	471	416	520	549	494	598	627	572	676	705	650
155	100	129	242	187	216	77	22	51	389	334	363	476	421	450	311	256	285	623	568	597	710	655	684	545	490	519
102	128	154	189	215	241	24	50	76	336	362	388	423	449	475	258	284	310	570	596	622	657	683	709	492	518	544
127	156	101	214	243	188	49	78	23	361	390	335	448	477	422	283	312	257	595	624	569	682	711	656	517	546	491
464	409	438	299	244	273	386	331	360	725	670	699	560	505	534	647	592	621	230	175	204	65	10	39	152	97	126
411	437	463	246	272	298	333	359	385	672	698	724	507	533	559	594	620	646	177	203	229	12	38	64	99	125	151
436	465	410	271	300	245	358	387	332	697	726	671	532	561	506	619	648	593	202	231	176	37	66	11	124	153	98
305	250	279	383	328	357	461	406	435	566	511	540	644	589	618	722	667	696	71	16	45	149	94	123	227	172	201
252	278	304	330	356	382	408	434	460	513	539	565	591	617	643	669	695	721	18	44	70	96	122	148	174	200	226
277	306	251	355	384	329	433	462	407	538	567	512	616	645	590	694	723	668	43	72	17	121	150	95	199	228	173
380	325	354	467	412	441	302	247	276	641	586	615	728	673	702	563	508	537	146	91	120	233	178	207	68	13	42
327	353	379	414	440	466	249	275	301	588	614	640	675	701	727	510	536	562	93	119	145	180	206	232	15	41	67
352	381	326	439	468	413	274	303	248	613	642	587	700	729	674	535	564	509	118	147	92	205	234	179	40	69	14

* 9 7 8 1 4 5 6 3 5 4 2 2 0 *